Is There a Balm in Black America? Perspectives on HIV/AIDS in the African-American Community

By Pamela H. Payne Foster, MD, MPH

With Contributions by Representative Laura Hall and Roland Barksdale-Hall

Foreword by Reverend Herbert Daughtry, National Presiding Minister of the House of the Lord Churches

Published by:
AframSouth, Inc.,
Montgomery, AL

The publication is designed to provide accurate and authoritative information in regards to the subject matter covered. If specific medical advice or expert assistance is required, the services of a competent professional person should be sought.

 ISBN # 978-1-4303-1670-1

Dedication

This book is dedicated to all those who labor on the front lines in the fight to decrease and eliminate HIV/AIDS in the African-American community and especially to the memories of the late Mr. Ellis Moorer and Preston Wilcox, both trendsetters in their respective fields, both passionate about their work, and both an asset to the community.

Acknowledgements

This book would not be possible without the support and encouragement of many. I first have to acknowledge the love and support of my husband, Bill Foster, who also shares my passion for issues that uplift and encourage our people. He also has allowed me the room and freedom to work, often at the expense of time with him, in my professional life and ministry.

I also acknowledge the love and support of my family, particularly my mother and father, Willie and Nina Payne, and my brother, Nelson Payne, from whom I share personal stories. In sharing my personal stories, I transected into their personal lives. I pray they understand the higher purpose that I attempted in expressing the effect of their life on mine.

I also want to thank the entire staff at the Institute for Rural Health Research at The University of Alabama, especially Dr. John C. Higginbotham and Leslie Zganjar and Mitch Shelton, who assisted in the editing and formatting of this manuscript. The staff always supports me in my personal and professional development and for that I am grateful. I also want to thank the administration, faculty and staff at Tuskegee University, which served as the inspiration for this work. Much of the work was delivered in speeches and writings while I worked in the Tuskegee University-University of Alabama Center of Excellence, Partnerships, Outreach, Research, and Training (EXPORT) Project, and was partially supported by the National Center for Minority Health and Health Disparities within the National Institutes of Health. Some of the reprints appear in the *Journal of Health Care for the Poor and Underserved*, the *Health Reporter* – Frontiers International Foundation Newsletter, and the *Montgomery Advertiser*.

I especially thank my final editor Mr. Roland Barksdale-Hall, a historian, genealogist, and author, who was more than an editor; who through the process gave invaluable advice and support in making this project a reality. I also thank him for his role in coordinating a National HIV/AIDS Book Project to donate books free of charge to community-based organizations. He is also a contributor to this project. I also thank the other contributor to the project, Representative Laura Hall of Huntsville, Alabama. Laura Hall, who represents the best of legislators, both in Alabama and in the nation, who are committed to eliminating this disease in the Black community.

Special thanks to Dr. Debra Henry Cooper, a medical classmate, and Dr. Larry Spruill, a friend of mine and my husband, who provided the design and the photo for the cover of this book, respectively. Their artistic and creative spirits as well as their business advice, is always appreciated. Special thanks also to Randy Jones and Anthony Merriweather of the HIV/AIDS Division of the Alabama Department of Public Health for their review of Alabama HIV/AIDS statistics and surveillance data.

There are countless friends and ministers who have encouraged me in my spiritual walk and believed in me over the years: Pastor Dr. Joyce Guillory, Minister April Johnson, Minister Vincine Brown, Minister Larry and Jackie Myles, Minister Dr. Faye Barclay-Shell, Dr. Valerie Clemons Weathersby, the "Long Island crew" including Carrie Alexander, Ora and Harold Waters,

Pastor Etta and Robert Weaver, Tom Ed and Mozelle Purifoy, Nancy and Henry Lee, Norma Jones, and Mrs. Tucker and godchildren Jessica and Erica.

I also thank all the spiritual leaders of all the congregations that I have belonged to, especially, the National City Christian Church in Washington, D.C., previously led by Rev. Alvin O'Neal Jackson, and the House of the Lord Church, Brooklyn, N.Y., led by Pastor Herbert Daughtry, and my current church, St. James Missionary Baptist Church in Montgomery, Alabama, led by Rev. Lee B. Walker, Jr. I especially thank Rev. Daughtry, who wrote the Foreword for the book.

I also want to thank the many mentors who helped develop me in my professional and, more importantly, personal path, most especially Sister Mary Carl at Xavier University, William Stone and Henry Moses both at Meharry Medical College while a graduate and medical student, Kathie Westpheling at the Association of Clinicians for the Underserved (ACU) previously with the American Medical Student Association (AMSA), Dr. Denice Cora Bramble at Children's Hospital in Washington, D.C. previously at George Washington University; and Dr. Dorothy Lane, my Preventive Medicine/Public Health Residency Director and colleague at the State University of New York at Stony Brook (SUNYSB). I also want to thank Dr. Francis Brisbane of SUNYSB, and Gwen Lipscomb, Director of the Office of Minority Health at the Alabama Department of Public Health.

Lastly, I want to thank three people who served as the inspiration and title for this book. Mr. Ellis Moorer, a former colleague at Tuskegee University and I spent many hours presenting HIV/AIDS 101 sessions throughout the state of Alabama. What made Ellis an excellent health educator was that he was totally consumed with passion for what he was doing. He was totally connected to his spiritual self and used that "power" to transform and heal his community. Additionally, special thanks go to Preston Wilcox, founder of

African American Families, Inc. (AFRAM), which serves as the model for AFRAMSouth, Inc. which will be the nonprofit vehicle for the National HIV/AIDS Book Project fueled by all proceeds of this book.

I thank Reverend Barbara Clemons for the inspiration for the title of the book. She gave a sermon at the Morning Prayer Breakfast of the Alabama Democratic Convention and I was sitting in the audience. She preached a sermon about Dennis Rodman, one of National Basketball Association's premier players who distinguished himself as the "master rebounder". She advised the crowd to be like Rodney, relentless in their "rebound" and in their struggle for control of the vote in the

November 2006 election. She asked us during the event, "Is there a balm in America?" I left renewed with purpose not only for the election, but with a title for this book. I hope the spirits of Ellis Moorer and Preston Wilcox encompass everything that I tried to accomplish with this book – to heal and promote good health in the struggle against HIV/AIDS.

Pamela Payne Foster, MD, MPH

Table of Contents

Foreword

This book is timely in addressing the epidemic of HIV/AIDS through a spiritual lens, particularly in the African American community. Our ministry, which is holistic, ministers not only to the soul but to the whole person - politically, economically, socially, and historically from an Afrocentric perspective and also from a health perspective.

So when the AIDS epidemic first hit our community of Brooklyn, New York in the early

1980s, our ministry discerned early on that we had a major problem on our hand. It was also a session by Dr. Benny Primm, noted national HIV/AIDS physician and leader, that really persuaded me to get myself and my congregation more involved. I became a founding member of the Black Commission on AIDS, an organization that tried to put policies in place to address the disease in the Black community.

Early in the epidemic, our church started Project Enlightenment, a program designed to educate pastors and churches about the epidemic and how to prevent it. It is ironic that the coordinator of Project Enlightenment is now the husband of the author. But perhaps the greatest test for our ministry came from one family. The McConnells had four family members, three daughters and one granddaughter, who all died from this disease. The pain and grieving suffered by the family and the congregation are forever seared in our memories.

The pain of a pastor conducting consecutive funerals for four women who died too early and too devastating a death seems cruel. It certainly shouldn't happen to others. Hopefully, the message

"Is There a Balm in Black America?"

of this book will resonate with the community. And its purpose will not be in vain.

Reverend Herbert Daughtry, National Presiding Minister, The House of the Lord Churches and Founding Member, National Black AIDS Commission

Introduction

A colleague in public health once said, "If we were doing everything we needed to do, our communities would be healthier." Why is it that African-Americans have some of the worst health statistics compared to other racial/ethnic groups? As a preventive medicine/public health physician, I have spent my whole career addressing this question. I have worked in a variety of settings: academia, community, clinical, and in large metropolitan cities, as well as rural communities. Although some things have changed, many of the themes remain the same: it is unclear exactly what contributes to these differences or disparities in health. Despite this enigma, the more I work in this area the more I know that one approach by one single discipline will not provide the answer.

When the Human Immunodeficiency Virus/Acquired Immunodeficiency Syndrome (HIV/AIDS) epidemic first evolved, I was a graduate student in Biomedical Sciences. As a part of my training, I had to complete both written and oral compulsory tests. Although my area of research was Sickle Cell Anemia, during the testing a broad spectrum of diseases was fair game for questioning. Someone asked me very detailed questions about HIV and AIDS and I stumbled over the questions.

Like many students locked up in a laboratory and focused on my specific area of study, I had not read much outside of the medical literature and certainly not public media. Little did I know that the initial ignorance on my part would evolve into my own personal passion for eliminating HIV/AIDS worldwide, particularly in the African-American community.

"Is There a Balm in Black America?"

As I have ventured from my natural training of public health, I have learned that spirituality and bioethical issues can augment public health strategies. It is in this vein that I attempt to present very nontraditional, contemporary issues that I feel may add to the current thinking in this field and apply them to the current HIV/AIDS epidemic in African-Americans. This book is a collection of essays, some published, many unpublished, and talks that I have given in very broad settings including churches, conferences, and social events for a broad audience.

Although I think that scientific and political communities are important components of solutions in eliminating this disease in the African-American community, it is the broader African-American community that is the primary target, and the Black Church certainly should be at the top of the list of organized institutions that should play a major role. I hope that this book adds to the dialogue and provides new ways of thinking about more comprehensive approaches to affecting this disease in order to answer the question, "Is there a Balm in Black America?"

African American Heritage and Healing

In Search of My Gullah Roots

When I was a child, I spake as a child, I understood as a child, I thought as a child; but when I became a man, I put away childish things. Corinthians 13:11 (NIV)

I am a Geetchie. White folks call it Gullah. As a child I always knew and recognized that I was a geetchie. I used to visit my great-grandmother in St. Helena Island, South Carolina, in the summer when I was younger and was struck by several things. First, it was always hot! Not just regular hot, but hot, humid and swampy. This was also evidenced by the enormous mosquitoes that devoured on our virgin skin. Secondly, people called my great-grandmother by her nickname "Titi." As a child I found this highly amusing because as far back as I can remember she was very old and spoke in a very "strange" dialect.

As an adult I now know that those child images of my heritage have so much more meaning. I have been begun the study of my family history through genealogy. Although many of our families have some oral history, genealogy is the systematic documentation of our history. African American genealogy is unique because we must also apply it to the context of white American history, as our history is interwoven through slavery, which is interwoven with white slave masters. Therefore, many of us begin to really learn or re-learn the **true** history of America during the time of slavery.

I have since learned that the coastal islands of South Carolina were unique during slavery times for several reasons. First, the geography of islands made them very isolated from the United States mainland. Because of this isolation, the only way to get to many of the islands was by boat or ferry. The hot, humid climate of this swamp area was advantageous for growing crops such as rice; however, many of the slave owners had no experience with this crop. It is believed that they relied heavily on the knowledge and skills of the enslaved Africans who were masters of the crop.

The weather of the islands was a disadvantage for the slaveholders because it was a perfect conduit for mosquito-borne diseases such as yellow fever and malaria, which the slaveholders could not physically withstand, forcing many of them to leave the slaves unattended. These facts combined to leave a unique group of slaves who were more independent than the average slave and able to retain much of their African culture, including language. Additionally, according to census records in the mid 1850s, many of my ancestors were unique in that they were able to read and write; some were even land owners. This may be in part to the establishment of the one of the first schools for the newly freed slaves on the island, the famous and historic, Penn Center, which based its curriculum on Booker T. Washington's philosophies of industrialism, education, and land preservation.

Pamela H. Payne-Foster

I originally got into genealogy by accident. I was assigned to be the Chair of the Reparations Committee at my church. My pastor, the Reverend Herbert Daughtry, a long-time community activist and social justice advocate, has often stood alone among Black Christian pastors in both domestic and international issues of social justice. It is interesting that even though one of the greatest examples of social justice of the Civil Rights Movement of the 1960s was birthed and supported by the black Christian community, many of the modern and current social justice issues have not been embraced by the African American church.

In fact, the church has either been silent or responds negatively to it. In fact, in particular, political or African-centered themes are not supported. For example, very few black churches are involved in the current Reparations Movement or other social issues such as global poverty and economics such as debt cancellation in Third World countries, particularly Africa, war, fair working wages, incarceration and death sentencing issues, and gun control.

When my pastor appointed me to leadership, I did not know a lot about the current

Reparations Movement in this country. I had been fortunate to be in a meeting with Bishop Desmond Tutu, Former Secretary General South African Council of Churches in South Africa earlier in 1997 and talked about their Truth and Reconciliation work at the beginning of their post-apartheid movement. I never will forget Bishop Tutu saying, "We think this process will heal our wounds, and that will put us way ahead of you African-Americans who haven't even gotten your forty acres or a mule."

I quickly got immersed in the Reparations Movement and learned that much progress had occurred. There was the recent victory at the International meeting of the United Nations where delegates from all over the African Diaspora and their supporters had gotten approval for wording that declared the Transatlantic Slave Trade a "crime against humanity." The actual wording opened the door for legal action against institutions including governments, academic institutions, and corporations.

Around the same time a young lawyer, Deidre Farmer-Paellman, conducted research and uncovered documents that tied a large corporation,

Aetna Insurance, to the transatlantic slave trade. Shortly after, many small cities and municipalities introduced legislation or resolutions in support of reparations for African Americans. Additionally, Congressman John Conyers (D–MI) introduced Bill HR 40 as a national reparations bill. Many long-time organizations that supported reparations such as the December 12th movement, National Black United Front (which Daughtry helped found) and the National Coalition of Blacks for Reparations in America was being joined by more mainstream organizations such as the National Association of

Colored People (NAACP) and the Southern Christian Leadership Council (SCLC).

Reverend Daughtry saw the clear connection between the reparations struggle and genealogy. If you account for your documented heritage, especially the connection with white slaveholders, descendents of slavery might be able to have evidence for compensation or reparations. But the strength of genealogy is even greater than that. The knowledge of who we are as a people, where we came from, our faith, and our great strengths and weaknesses all point us to our purpose in life. We should study and know our history in order to constantly remind ourselves and educate others on our greatness. It is who God created us to be!

Our purpose should also be clear on who we are as a people in maintaining healthy lifestyles and good health. This purpose should translate in all diseases including HIV/AIDS. God has promised us an abundant life. We must be clear on what we have to do to walk in it. God will provide a way. He will heal our land! Amen.

Restoring A Deferred Dream

Uphold me according unto thy word, that I may live: and let me not be ashamed of my hope.
Psalm 119:116 (NIV)

I recently saw a special on the Public Broadcasting System (PBS) in which the producers interviewed a group of young children years old – black and white – and each about six years old – and asked them what their dreams and hopes were. There was no difference between them. They all had great hope; many said they even aspired to be president. Then the producers interviewed the same group five to 10 years later, in their adolescence, and discovered something interesting. The black children did not seem to maintain the same optimism and hope that they had expressed a few years earlier. What factors contributed to the dramatic change?

I suspect that many of the Kwanzaa principles, which are based on spiritual and social principles, could offer some solutions. Kwanzaa was founded in 1966 by Dr. Ron Karenga and is the only African American major holiday devoted to the development and growth of African Americans in the

United States. I had opportunity to hear Dr. Karenga speak about how he came to found Kwanzaa. The idea for Kwanzaa came to him during a crisis as a graduate student during the 1960s.

He felt that classification of the United States as a melting pot was wrong. He believed that blacks needed to come to the realization that we are not like others, we are unique and that we should embrace that and build on it. Furthermore, we needed to go back to the old time ways of our ancestors to sustain us and encourage us as a people. In other words, we should embrace the old "Way of Life".

I think these principles and Karenga's purpose for the holiday fit into what we now know as "Healthy Lifestyle /Behaviors," which are important in eliminating health disparities in particular diseases like cancer, cardiovascular disease, diabetes, HIV/AIDS, and infant mortality. Let's look at HIV/AIDS in the African-American community as an example of how the Kwanzaa principles could be applied. Let us look at how gay white men and their fight against AIDS in the early years of the epidemic in the early 1980-90s.

Seven principles are associated with Kwanzaa. It is African-based (Swahili) and the order is important. The first principle is **Umoja – Unity** – the entire African-American community should be working to eliminate HIV/AIDS. Gay white men were unified in the fight against HIV/AIDS; they saw it as a public health emergency and the Number 1 disease (killer) in their community. Currently, of the new cases of HIV, African-Americans represent more than 70 percent of all new cases and Black women represent over 70 per cent of new cases in women in Alabama in 2005. We in the African-American community need to agree that HIV/AIDS is a problem and that all should be involved in its solution.

The second principle is **Kujichagulia - Self determination** – Gay white men took HIV/AIDS on for themselves, didn't wait on government or anyone else to assist or help them. If we remember the Vice Presidential debate in the 2000 election, neither candidate had good knowledge of the HIV/AIDS epidemic in the African American community. Either the government has no idea or doesn't care about the seriousness of the epidemic. We need to act and not wait on the government.

The third principle is **Ujima–Collective work and responsibility** –It is going to take the whole village – church, schools, parents, etc. to solve this epidemic in our community. If we use gay white men as an example, they worked as a unit, collectively and responsively around the HIV/AIDS epidemic. They researched all aspects of the disease, they organized around all areas, they served as strong advocates for those affected, and they lobbied strong for adequate screening, testing, and treatment. For example, they lobbied to federal agencies and pharmaceutical companies for faster approval of AIDS drugs and they also lobbied for new governmental funding streams for clinical treatment such as the Ryan White Comprehensive Act.

The fourth principle is **Ujamaa– Cooperative Economics** – Any organization or movement needs money to get the word out, to get people educated, tested, evaluated, and conducting research, etc. For most new funding streams for

HIV/AIDS – gay white men were intimately involved in creating funds. In addition to creation of funding on the federal level, many of the gay white activists raised millions of dollars through nonprofit organizations to assist their movement.

The fifth principle is **Nia – purpose** – The African-American community needs to be clear on the AIDS issue and the severity of the epidemic. Gay white men exhibited a clear purpose in the elimination of the disease in their community. It was a focused effort and became the larger purpose of the community. Their efforts helped drastically decrease the HIV/AIDS rate in their community.

The sixth principle is **Kuumba – creativity** – In the early 1980s through the 1990s, gay white men used creative methods to decrease HIV/AIDS. One such technique was providing AIDS 101 sessions at bathhouses and passing out condoms at nightclubs and adult specialty book and video stores.

The last principle is **Imani – faith** – It will require a spiritual awakening that encourages people to recommit to moral standards to restore our community and prevent HIV/AIDS. As we move forward, I believe these same principles can answer so many of our communities needs to fulfill healthy lifestyles and eliminate health disparities. We know for ourselves that if we didn't have faith in our lives, we would never have survived the African holocaust. We thank God for the principle of Imani. Amen.

Discussion and Reflection Questions – Introduction

1. What is genealogy and what is the value of it?

2. In the study of genealogy, one often starts with oneself. Discuss or write a short autobiography that you would leave as a legacy to your ancestors on some of the facts of your life, as well as a depiction of who you are as a person.
3. Discuss the seven principles of Kwanzaa and how each principle could be tied to providing solutions for decreasing HIV/AIDS in the African-American community. (Some examples are given in essay " A Dream

Deferred.") Please refer to examples and give other examples.

Chapter One
The New Face of HIV/AIDS

Is AIDS a New Black Holocaust?

Give strong drink to him who is perishing, and wine to those who are bitter of heart. Let him drink and forget his poverty, and remember his misery no more. Open your mouth for the speechless, in the cause of all who are appointed to die. Open your mouth, judge righteously, and plead the cause of the poor and needy. Proverbs 31: 6-9.

Recently, I was at an international conference focused on cultural competence issues in substance abuse, addiction, and HIV/AIDS and the presenter put up a map of the cumulative rates of HIV/AIDS throughout the world. He stated that although currently many developed countries such as Europe, North America, and China were experiencing epidemic proportions of the disease, it was clear that Sub-Saharan Africa, the Caribbean, as well as African-Americans were carrying the heaviest burden of the disease.

He also said something very poignant. The presenter said: "If the common thread is that those who are carrying the heaviest burden of the disease are of African descent, we may have another "Holocaust" on our hands." His words hit me like a ton of bricks and are the take-home message that I want to leave with the community.

When most Americans think of the word holocaust we think of the terrible "Jewish Holocaust," where millions of European Jews were systematically murdered in order to exterminate their presence. I had the opportunity to visit the National Holocaust Museum in Washington, D.C., and the experience was a harrowing one. I learned of how insidious and systematic Hitler and his followers were at work at so many levels in their extermination of a group of people.

Examples included destruction of literary writings and intellectual properties of Jewish scholars, the cruelty of the concentration camps, including the cutting of hair, and collection of personal property of those encamped in order to strip them of their humanity. I also learned of the medically cruel experimentation of Jewish persons without their consent. The museum's existence is a constant reminder not only the atrocities that happened but is also is a constant reminder that such occurrences should never occur again.

But the word "Holocaust" has significant meaning to many in the African- American community who remembers and will never forget the institutionalized and systematic Transatlantic Slave Trade that displaced millions of Africans from their homeland and resulted in the deaths of many. It is unfortunate that we do not have a perpetual institution that can constantly remind us and keep us focused on never letting it happen again.

Pamela H. Payne-Foster

During the 25-year course of the current HIV/AIDS epidemic globally, I believe that many people of the African Diaspora have not been proactive in addressing this epidemic, and 25 years later we are paying dearly with the lives of millions of people. Whenever the lives of millions of people are taken away when it could have been prevented it is a Holocaust. And so it is with this disease.

I cry inside every time I hear of any Holocaust or injustice that ignores a serious problem. Modern day Holocausts would include the lack of adequate housing, employment, and health care for the poor, especially people of color. The Brown and Black complexion of the modern American public educational system, criminal justice system, and the industrialized prison system looks eerily like a throwback to slavery and has reached epidemic proportions for many men and women of color. It could be described as a Holocaust. Institutionalized racism was the engine that drove slavery 400 years ago and is most likely the culprit behind discrimination today.

The thread of racism that connects these modern day Holocausts certainly does not escape the disease that we call HIV/AIDS. The lives of African-Americans, Caribbeans, and Africans who died from this disease, the families they left behind, and the devastation that such large burdens leave on our communities is certainly is symbolic of a Holocaust.

African-Americans have to be at the forefront of making this wrong right. This will mean that first we will have to make the issue of global HIV/AIDS our Number 1 priority. A Holocaust is an emergency situation and it deserves the undivided attention of the communities affected.

Secondly, we will have to be united in our strategies for eliminating this disease. People with diverse lifestyles, occupations, socioeconomic levels, gender, sexual orientation, geographical residences – those in the city, those in the country or rural areas, those in suburban areas-will have to come together around fundamental prevention and treatment strategies. For example, because most HIV/AIDS activists know that one-third of persons infected with HIV are unaware of their status, I believe most would agree that increasing testing so that the majority of African-Americans know their HIV status should be a fundamental strategy. The particulars of how to accomplish this strategy for different subpopulations of the African-American community might vary, but we should be united on this fundamental strategy.

Most HIV/AIDS activists also know that strong cultural barriers about homosexuality have to be addressed in order to develop effective prevention strategies for men who have sex with men. This can be accomplished in a number of ways, but the community has to be united in the fact that HIV/AIDS cannot be eliminated in the African-American community until we can agree that homophobia is a barrier in prevention strategies and must be eliminated in order to eliminate this disease.

HIV/AIDS is hemorrhaging our community. We must begin to stop the hemorrhage at all points: globally and throughout the African Diaspora; at critical barrier points in order to develop prevention strategies that will work in our communities; with increased testing and access to care and treatment; and with adequate resources and research in order to understand where hemorrhages are and how best to stop them.

Pamela H. Payne-Foster

I long and pray for the day when I can say. "The HIV/AIDS Holocaust is finally over; but I will never forget." AMEN!

Discussion and Reflection Questions – Chapter One

1. Is there a connection between the HIV/AIDS epidemic in the African-American community, the epidemic in Sub-Saharan Africa, and the Caribbean? If so, what is it?
2. How should we prioritize our efforts to address the global HIV/AIDS issues across the African Diaspora?
3. Do you think the African-American family has the power to eliminate HIV/AIDS within the community?

4. What do you think are the top three barriers within the African-American community in addressing HIV/AIDS?
5. What strategies would you think the African-American community should use in addressing the barriers in Question #3?

Chapter Two

God Will Heal the Land
"A New Hope" *

Which hope we have as an anchor of the soul, both sure and steadfast, and which entereth into that within the veil? Hebrews 6:19 (NIV)

*** Appeared in the *Journal of Health Care for the Poor and Underserved,* August 2004- Letter to the Editor**

As I walked with a large procession through the financial district of New York City in the fall of 2003, I could barely feel on my face the cool wind blowing softly through the tall buildings. During the beats of the African drums, I thought I heard the soft whispers of my ancestors crying out in pain and despair from their centuries-old burial ground.

I was one of several thousand participants in a procession for the reburial of the bones of four hundred Africans discovered in lower Manhattan in 1991 in what had been the borough's 18th century African Burial Ground. After more than a decade of study and analysis at Howard University in Washington, DC, a week-long schedule of spiritual and sacred events held throughout the Northeastern seaboard culminated with activities in New York.

The remains were being returned to their original resting place.

The discovery was historically significant for many reasons. First, their discovery reminded the American people that enslavement of Africans in this country was not strictly a Southern phenomenon. The bones date back to the early 1700s. The early settlers of colonial New York used enslaved Africans to help build the new land; these Africans built the very wall protecting the financial district after which Wall Street is named. Second, the bones bore signs of the cruelty and abuse often associated with slavery. There was evidence of fractures, other skeletal lesions, and weakened bones due to malnutrition. In fact, many of the bones were of women and children.

During a funeral in the Deep South of Alabama several months later, I recalled the feeling I had on that cool fall day in New York. In Alabama, I was at the funeral of Mr. Ernest Herndon, the last survivor of the infamous United States Public Health Service Syphilis Study (also known as the Tuskegee Syphilis Study). As attendees gave testimonies, I tried to imagine the pain and agony that Mr. Herndon and his family endured for the long years during and after the study.

Mr. Herndon was one of over six hundred Black males who unknowingly participated in a study that deliberately withheld treatment from syphilitic participants. For forty years, government scientists observed and documented how syphilis ravaged the human body. Even when penicillin, a much better treatment for syphilis than had previously been previously available, became more widely available in the 1950s, treatment was denied to the men.

Injustices experienced by the New York ancestors and the Syphilis Study participants are tied together by the threads of exploitation and racism, both of which are sometimes as evident in silence as in non-silence. Over the Syphilis study's long duration, researchers published numerous publications in professional medical journals, but stayed silent about the ethical violations of the study. Throughout the centuries of legal slavery on these shores, many Americans who were not slaveholders never the less stayed deafeningly silent about the institution and injustices of slavery.

Such racial injustices do a disservice not only to the victims, but also to society as a whole. In order to address injustices, we must first begin to recognize them. In 1997, then President Bill Clinton both recognized and addressed the injustices of the United States Public Health Service Syphilis Study through his formal apology for the role of the federal government in carrying it out. It was the first step in the right direction and one that could serve as a model.

I attended Mr. Herndon's funeral in Alabama as a new faculty member at the Tuskegee University

National Center for Bioethics in Research and Health Care. The Center was established in 1999 with a primary goal: to examine bioethical issues in health that affect underserved communities, particularly African-American communities. The Center comes on the scene at a time in this country's history when health disparities between African-Americans and whites are widening.

Cardiovascular disease, cancer, diabetes, and HIV/AIDS are devastating Black communities. Many theories have emerged to explain these disparities. Any satisfactory theory must include how hundreds of years of racial inequalities have fostered them. Without understanding the roles that racism and other injustices have played and currently play, and acting to correct these injustices, we can never eliminate these health disparities.

At the funeral, Attorney Fred Gray, who represented the survivors of the study in their lawsuit against the United States government, said that Mr. Herndon's death was "the end of an era." His words rang true in my spirit. The funeral reminded me of the brutal history, some of it in the annals of science and research, of African-Americans in this country.

The occasion also reminded me that we all have much work to do on such matters as trust and accountability, if we are to heal and repair historical and past damages that hinder the reduction of gross health disparities in this country.

At the funeral, I heard the soft whispers of my ancestors calling me to renew my purpose and refurbish my hopes. As a health care provider, I am working both to recognize and address injustices in health. I encourage others in a variety of health fields (e.g., dentists, nutritionists, nurses, social workers, physicians, pharmacists, and health educators) who can make a difference in eliminating health disparities to both recognize and act on these issues. Together we can begin to make a difference for millions. I will keep listening for the voices of my ancestors as I strive to work in the health disparities/social justice arena. We owe as much in memory of the hardships that so many bore!

Discussion and Reflection Questions – Chapter Two

1. *The Bible* and other spiritual books give numerous examples of healing by God, Jesus, and other prophets. Give a few examples with scriptural references. (For example, one of my favorites is II Corinthians 3: 17 – *In Christ, there is liberty!*)
2. Are there any modern day examples within your own life of miracles of healing?

3. Do you personally know any persons with HIV/AIDS? Do you think that God should heal them and, if so, in what ways?
4. Discuss the statement, "God can heal the land" and what it means for you and the current HIV/AIDS epidemic in the African American community. (Refer to Appendix A on how the HIV/AIDS epidemic is currently disproportionately affecting the African-American community.

Chapter Three
Intimacy and Relationships in HIV/AIDS

Jada, R Kelly and E. Lynn Harris: Lessons for Healthy Relationships for Black Women in 2006

Blessed be the God and Father of our Lord Jesus Christ, who hath blessed us with all spiritual blessings in heavenly places in Christ. Ephesians 1:3 - KJV

When HIV/AIDS came on scene in the United States in the1980s, the major risk populations for contracting the disease were gay white men and intravenous drug users (IVDU). There were small populations of other risk populations, including those who had frequent transfusions, such as hemophiliacs and other international populations such as Haitians, but the disease was largely restricted to those big-risk categories. Heterosexual transmission to women was virtually unheard of in this country.

However, if you paid attention to what was happening with the HIV/AIDS epidemic worldwide or globally, particularly in Sub-Saharan Africa, heterosexual women became infected in large numbers, which would provide us with some clues for what could happen with the disease in the United States some 25 years later. At the time when the disease looked different in Africa than in the United States, the question was why?

The first thought was that the virus was different in the United States and in Africa – one type transmitted homosexually, the other type transmitted heterosexually. The other train of thought was that of the environment playing a role in how the disease looked in Africa; that of Africa being a developing continent where malnutrition and poverty, although scattered throughout the United States is proportionally far greater and significantly worse than in America.

The global transmission of the disease did indeed provide a foreshadowing of increased rates of transmission of the disease in people of color and in women of color. In 2004, the Centers for Disease Control and Prevention reported that 50 percent of all new cases of HIV reported were in American-Americans. In 2004, of women affected by HIV, 70 percent of new cases were African-American. HIV/AIDS caseworkers in Alabama report that currently a typical case of a newly diagnosed HIV+

person is a 40-year-old married African-American female.

What could be some of the reasons for increased rates in women? Certainly a lack of negotiation skills and self-esteem in the use of condoms in their male spouses could be a factor. Additionally, increased rates in males in prisons, which contain disproportionate numbers of African-Americans, may increase rates in women who have sexual contact with male inmates upon their release. Lastly, there is the newly reported reality of the secret lives of men who have sex with men, who don't identify themselves as gay and who are engaging in heterosexual relations.

These relationships can occur across the community – in married women or women who think they are in monogamous relationships and in young females in high school and college who engage in sexual relations with men who have sex with men on the "Down Low". Lastly, economics may play a role for women of color who disproportionately have decreased access to health care.

I believe that these theories come down to one common denominator in our risk as women in contracting HIV/AIDS: Relationships, Relationships, Relationships. I want to lift up three names as models for prevention of HIV/AIDS by women in the 21st century: E Lynn Harris, R. Kelly, and Jada Pickett Smith. All are celebrities, entertainers, and famous. What does each have to say about how we should be conducting relationships in the era of HIV/AIDS?

Back in the day, when I was dating in college, the sexual revolution was in its hey day.

You might sleep with a man after having just met him at a party or the club. You didn't ask any questions and you didn't feel bad about it. It was the way it was. We had condoms in my day, but our concern was more about not getting pregnant. So some of us who were on the pill did not necessarily use condoms. We were letting it all hang out. We were experimenting. We were having fun. That was in my day. The thought that any of the guys I slept with during that time were sleeping with men never even crossed my mind.

In fact, I was very naïve, until I read the books *Invisible Life* and *Just as I Am* by E. Lynn Harris. Mr. Harris described, in often graphic terms, the secret lives of men who were bisexual and leading double lives. One character was a very buff football player; another a professional. It shattered my misconceptions about the stereotypical gay male – effeminate and obviously gay. On the contrary, his characters were masculine and very closeted.

I read the books during a time in my life where as a single professional living in large metropolitan cities like Atlanta, Washington, D.C.

and Long Island New York. I began to recognize large, gay communities with Black men, many closeted. Therefore, I could not date like I used to; making assumptions, thereby putting myself at risk. It opened my eyes to the whole idea of alternate lifestyles, which are definitely a reality in 2007 in the era of HIV/AIDS.

Which leads me to a second person who has a lesson for how women should handle relationship in this current era of AIDS. R Kelly is a rhythm and blues singer. Although a talented entertainer, he is not without controversy in his own personal life. But his recent video documentary, "In the Closet," offers insight into the complicated lives of many. The video documentary spins a web of lies, deceit, and drama that is real in 2006. For women, it means operating in accordance with this reality. R Kelly adeptly weaves stories of living on the down low (male and female), adultery, prison life, and the party life, which all potentially increase the risk of HIV/AIDS. Both E Lynn and R Kelly show us why we should be asking questions of those we date, why we should use condoms consistently and correctly every time we have sex, and why we need to be careful how we mix alcohol because of its connection to risky sexual behavior.

But I think Jada Pickett Smith gives us the best model for how to avoid contracting HIV/AIDS. I recently saw a biography on her husband Will Smith. Before this documentary, I knew that Jada Pickett grew up in Baltimore, Maryland, and went to the same performing arts high school as Tupac Shakur. I have always been interested in biographies because if done right, they show patterns in life and critical events that move people to where they are.

Interestingly, Tupac Shakur's mother was a Black Panther/activist and eventually moved to Brooklyn, New York, where she and her son were members of my church when I lived in New York. My pastor tells the story of Afenyi and Tupac joining the church when Tupac was 8-years-old and the pastor asking him, "Young Man, what do you want to be when you grow up?" Tupac answered him, "A revolutionary." Powerful and prophetic words from such a young boy. Jada and Tupac were close friends, both interested in performing arts. Jada would eventually become a famous actor and dancer; Tupac a famous actor and rapper. Whether or not they dated is not known, but I can imagine that Tupac might have been interested. Yet Jada wasn't to go down that path.

The biography on Will Smith revealed that she instead dated the basketball player, Grant Hill, currently married to Tamia, another famous singer. She eventually endured a hard break up with Grant Hill, which led her to start dating Will Smith and eventually to marry him. I thought about Jada's taste in men and the decisions she has made in determining suitable mates for dating and marrying.

It seems to me that she has very discriminating taste. Both Will Smith and Grant Hill seem to be upstanding and responsible Black men. That is the impression they give and I think she certainly could not have gone wrong with either one. Both are accomplished in their fields, are from solid family backgrounds, and appear to be family men. Often, when women reach the socioeconomic status and fame that Jada has, they say they cannot find quality Black men.

That is a myth. I remember before I met my husband, I was on the "ain't no good Black men around " syndrome and my husband, who at the time was just a friend, said to me in his quiet way, "Pam, all you need is one good Black man." In other words, look a little closer. You see, I had said things like too old, too fat, too short, too tall, not enough education, not a good enough job, too many children, etc. I was neglecting things like Godly man, a man of character, a man of principles.

Women, let us get serious about how we choose our partners. Every decision we make could be the difference between happiness and unhappiness, struggle and peace, and life and death. Amen.

Discussion and Reflection Questions – Chapter Three

1. What are some of the unsafe sexual practices that you know will cause HIV/AIDS?

2. Are you or any friends or relatives engaging in risky sexual behaviors that may put you at risk of HIV/AIDS? How do you feel about it?

3. Do you know anyone either in your family or a friend who is openly gay? How do you feel about it?

4. Do you suspect someone may be gay and is not openly disclosed their sexual status? How do you feel about it?

5. If you were gay would you be open? Why or why not?

6. Why do you think it is so difficult to discuss sexuality and homosexuality issues openly in the African-American community?

7. Discuss solutions in making subjects such as sexuality and gender orientation (homosexuality) more open in the African-American community.

Chapter Four

Prevention in the College Coed

"HIV/AIDS for the 21st Century College Coed"

When I was a child, I spake as a child, I understood as a child, I thought as a child; but when I became a man, I put away childish things. I Corinthians 13: 11 – KJV

While I was matriculating through college, I was curious; I studied a little. I partied a lot. I made many friends. I dated a lot of young men. I partied a little more. I grew spiritually. I sang in the gospel choir. I pledged Delta. I became a cheerleader. I studied a little more, and I partied some more. I was in a great party city - New Orleans - so how could I resist! It was the first time I was away from home and I drank alcohol for the first time, did a little drugs for the first time, and had sex for the first time – much like a College coed would in 2006.

However, at that time in the late 1970's and early 1980's, AIDS and HIV were not a reality. In 1981, when I graduated college, the world began to hear about this new disease, first characterized as a disease thought to be caused by a virus (HIV). I say thought to be because there are still some scientists that believe the virus is not the most important factor in causing disease, but that malnutrition and poverty are the rate limiting factors.

At that time, the disease occurred mainly in gay white men and intravenous drug users (IVDUs) in the United States; it did not penetrate personally into my world or the world of many other African Americans. But gradually over the years, we heard of the epidemic in other parts of the world, mainly sub-Saharan Africa and, for the first time, in heterosexual women.

Fast forward to 2006. The newest cases of the disease are predominately in African-Americans, both men and women. The South is the fastest growing region of disease growth. Small metropolitan areas and rural areas have joined the ranks of cities like New York, Los Angeles, San Francisco, Washington D.C., and Atlanta, including Charlotte, North Carolina., Montgomery, Alabama, Columbia, South Carolina, and Savannah, Georgia. In 2002, an epidemic of HIV/AIDS occurred in several Historically Black Colleges and Universities (HBCUs) in North Carolina. Many of those infected were Black males. It was the first documented HIV outbreak on a cluster of U.S. college campus. According to researchers, the main cause of increasing HIV infections in this cluster was young Black men who were having unprotected sex with other men but and not self-identify as gay or bisexual.

In a study targeted at students at affected colleges in North Carolina, 84 new cases of HIV were found, 73 of which were Black. Of those, 67 reported sex with men, and 27 of those also had female sex partners.

Although the college coed in 2006 is not much different than in my day, the risks of contracting HIV/AIDS are much greater. What do I think are the risks and issues and how can college students protect themselves?

1) Lack of knowledge. Most college students are very nonchalant about the risks of HIV/AIDS and are not knowledgeable about transmission of the disease or how to

adequately protect themselves. I think most have very superficial knowledge. They know what HIV and AIDS stand for; and they know a little about the virology. However, there are still many myths, misconceptions, and misinformation about the disease – particularly about ways the disease is transmitted and specifically how they can protect themselves. In my experience, the average college student remains unaware about:

x Harm reduction behaviors such as needle exchange programs and sterilization of drug apparatus works and the use of apparatus used to protect yourself in oral sex such as dental dams or saran wrap.

- Information about the correct and consistent use of condoms
- Knowledge about how to protect your baby if you are infected and are pregnant
- Specific knowledge about testing, what the test means, and how often and when you should get tested. In a college setting where intellect is the foundation of education, <u>all college students should be encouraged to know and learn as much as</u>

they can. The Centers for Disease Control and Prevention website www.cdc.gov has a wealth of information. Use it.

2) College students do drugs. This is not a new phenomenon. But drugs do cloud judgment, particularly in condom use – using condoms correctly and consistently. This includes all drug use: alcohol, prescription drugs, and marijuana, as well as harder drugs, such as cocaine, heroin, and methamphetamine. My advice to all college students in this day and age is to drink alcohol and take prescription drugs

responsibly and stay away from illegal drugs. Also, avoid other risky behaviors that can put you at risk of transmission of HIV such as sharing tattoo, piercing, or barber equipment.

3) Hip Hop/Gangsta rap-music videos. This form of music was not around in the 1970s when I was in school. Although it is natural for young people to be attracted to forms of music different than their parents, I think current hip hop music in the form of gangsta rap pushes the boundaries of creative free speech in the areas of sexuality and

violence. These behaviors, often subtle, sometimes not, get played out in real life. For example, last week in church, I saw a mother wearing a necklace with the word – Sexy. This was the mother, not the child. Times have changed where even ten to fifteen years ago that would have been unheard of. Songs like "Get your Freak on, Back it Up, etc." have explicit lyrics which often display sexual acts very openly in music videos. These images and lyrics have produced a culture that introduces sexuality to our children often and at earlier and earlier ages. The counter message of abstinence or responsible sex is minimal. My advice to the college coed in the 21st century is to wait for your sexual encounter, but if not, use a condom correctly and consistently.

4) Sexually transmitted diseases (STDs)/ Sexually transmitted infections (STIs)– These issues are not new; we had them when I was in college. But they have greatly increased since my time and they are implicated as co-infectors in the transmission of HIV/AIDS. They include

chlamydia, syphilis, gonorrhea, and venereal warts. It appears that they make the environment conducive for transmission of the virus. So the message is: use a condom correctly and consistently.

5) Homophobia – Men having sex with men is the Number 1 group infected with HIV in African-Americans. The study in North Carolina concerning HBCUs is probably not an isolated occurrence. It is highly probably that these behaviors are also occurring in other HBCUs across the nation. My experience at several HBCUs is that the

campuses are generally extremely hostile to alternative lifestyles, thus, forcing men to have sex with men in secret and fueling the disease in both men and women. A recent study at Morehouse College in Atlanta confirmed such attitudes. Encourage tolerance of alternative lifestyles. Do not make assumptions and know the HIV status of all sexual partners.

6) Lack of testing and knowledge of one's HIV status. With such high rates of HIV/AIDS in the Black community, particularly in young people who exhibit risky behaviors, every

college coed should be regularly tested for HIV and know their status. You should also know the status of all your sexual partners, be they male or female. Period. Because HIV can give no symptoms, many people are infected and don't know it and are unknowingly spreading the risk. Therefore, get tested, know your status, and make sure everyone you become intimate with gets tested and you know their status as well.

Lastly, my advice for the college coed is to encourage those who are living with HIV or AIDS. I am sure there are some persons currently at HBCUs who are HIV+ or have AIDS. There are many persons living successfully with HIV/AIDS. Some we may know about, others prefer to keep it confidential.

Our most famous African-American successfully living with HIV is Magic Johnson. How has he lived so long with virus, some 15 years? He said his secret is that he was diagnosed early and he takes his medicine consistently. He eats a nutritious diet and exercises frequently. He limits his stress, but most importantly he maintains and strengthens his spirituality. Magic Johnson provides a wonderful model for healing and good health in the land. Amen.

Maybe Jocelyn Elders was Right!

Now Lord, consider their threats and enable your servants to speak the word with great boldness.
Acts 4:29 (NIV)

HIV/AIDS is currently at epidemic proportions in the African-American community and is rapidly increasing in the rural South. For these reasons, we have been conducting a campaign to take HIV/AIDS education out into the Alabama Black Belt region. The region gets its name from the color of the soil in this area – black and rich and historically great for farming. The region also is home to some of the worse health statistics in the state and nation. As a part of our campaign, we visited a detention center where we conducted a HIV/AIDS presentation. The session included the use of media messages, a motivational speech by an infected individual, and a short HIV/AIDS 101 educational session that included a demonstration of how to correctly use a condom. The session was targeted to 350 young boys ages 12 to 17. According to the director of the center, most of the boys fit into high - risk categories for HIV/AIDS infection, so the session was greatly appreciated by most of the center administrators and staff and the young men.

Despite the acknowledgement of the need for straightforward HIV/AIDS education by the administration, one of the teachers became agitated during the presentation. She said, "There are 12 -year olds in the audience!" I scoffed at the denial. Although some might be abstinent, the likelihood is that many are not. At what age should we encourage and teach correct condom use?

Because there is no cure for HIV/AIDS, harm reduction and behavioral techniques, which have been proven to work in preventing HIV/AIDS, are vitally important. Certainly correct condom use is key in prevention efforts. Research indicates that correct condom use or condom use in general may be limited in certain communities such as among African-Americans and Latinos.

The question then becomes: At what age should public health encourage and teach correct use of condoms? Teenage years, certainly before youth become sexually active, and initiation of sexual activity - what age is that? Research indicates that some children initiate sexual activity as early as the age of 9.

Therefore, responsible sexual education, which includes correct condom use and which captures this small group of youth, should probably begin at the elementary school level. However, many in society are uncomfortable with beginning the very explicit conversations that HIV/AIDS prevention requires at this age. In fact, I have found from my own experiences in HIV/AIDS education to community groups that there is some level of discomfort associated with giving very straightforward sexually explicit information to even high school students and adults.

HIV/AIDS is at staggering proportions in the African-American community, particularly in persons aged 19 to 25. I remember when Dr. Jocelyn Elders, the former Surgeon General of the United States, talked frankly and openly about the need for early sexual education. Many in the public health community were ecstatic: finally, a public health official who had a pragmatic approach to public health. She dared to go where many public health officials, because they are political appointees, do not. She dealt head-on with very controversial issues about sexuality - the legalization of marijuana, condom use, reproductive rights around the early morning pill, and even masturbation.

Perhaps she went a little too far in some aspects, such as her comments about masturbation and legalizing marijuana. The controversial comments took the focus off her public health approaches and required her to defend herself politically. Despite her great support from many in the public health community who know she was ahead of her time, there were many others who condemned her. From my vantage point, watching in horror as HIV/AIDS rates continue to worsen for the African-American community, maybe Jocelyn Elders was right.

Discussion and Reflection Questions – Chapter Four

1. Why do you think the African-American college student is at particular risk from contracting HIV?
2. What are some of the risky behaviors that college students engage in that might be considered as "indirect" risk factors? (Hint: alcohol and other drugs can decrease the consistent and correct use of condoms and increase the risk of date rape)

3. Discuss the particular outbreak of HIV/AIDS in North Carolina Historically Black Colleges and Universities (HBCUs) and its implications for young African-American college students in your home town.
4. What roles should college administrators, faculty, students and the public health community play in ensuring that the risk of contracting HIV/AIDS on any college campus is decreased?
5. What solutions can you give for increasing diversity of sexual orientations and tolerating

of differing sexual orientations on a college campus?

Chapter Five❋

The Dis-ease of HIV/AIDS

The Dis-ease of HIV/AIDS: Implications for the African American Community*

For God hath not given us the spirit of fear; but of power, and of love, and of a sound mind.
2 Timothy 1:7 (NIV)

*** Appeared in the *Montgomery Advertiser*, February 13, 2006**

I have worked as a health professional in the fight against HIV/AIDS in the African American community since the early 1990s, and despite science's increased knowledge about the virus and its transmission, translation and internalization of this knowledge into prevention strategies in the Black community still remains a challenge. The prevention messages, which often rely on straightforward and direct discussions around obviously sensitive risk behaviors such as unprotected sex, substance abuse, homosexuality, bisexuality, incarceration rates, and not so obvious risk behaviors such as extramarital or other sexual relations outside boundaries of relationships, and the sex worker/prostitution industry, appear to be too uncomfortable for many in the community to seriously address in order to prevent this virus. But until this dis-ease is addressed, the current HIV/AIDS epidemic that currently plagues the African-American community will only get worse.

In 2004, 50 percent of all new cases of HIV/AIDS occurred in African-Americans. The majority of new cases of HIV/AIDS in women are in African-Americans. Most in the community have some knowledge of HIV/AIDS, - its modes of transmission and the increases in the African-American community - but apparently not enough knowledge to affect change. So what is preventing people from being well - informed and/or adjusting their behaviors in order to reduce their risk of infection?

I recently participated in a HIV/AIDS outreach tour targeting the Black Belt counties of Alabama, which is predominately rural, African-American and low-income. The tour involved using a multifaceted approach, including multimedia, college peer educators, HIV/AIDS consumer advocates, and well known entertainers and authors in a variety of settings, including universities, high schools and community town hall meetings.

Probably the most poignant recurrent theme on the tour was the elements of stigma, fear, and denial (SFD). I propose that these three elements are the real dis-eases that we must address in order to prevent future cases of HIV/AIDS. SFD may be even more pronounced in rural settings, because of the nature of rural areas where towns are small and everyone knows each other. This inherent close-knit nature of rural areas probably frightens people to either keep their risk behaviors or HIV status secret.

For example, we learned from antidotal accounts during the tour that many affected persons in rural areas avail themselves of testing and treatment services outside of their home county or state. Why? It is probably because of FEAR driven by the stigma of HIV/AIDS. Many affected persons might be afraid of community members knowing about their illness or their risk behaviors and fear rejection.

Many others lack even a desire to know their status. Additionally, the fear of breaches of the confidential release of information is a constant threat when they use services in their home towns or locally. Stories were told of confidential lists being distributed in towns and staff members divulging the HIV status of their clients.

The Black Belt HIV/AIDS Tour activities in which education and HIV testing were conducted were marketed by poster leaflets, radio advertisements, and use of nationally known personalities and entertainers. Despite these efforts, attendance at audience events such as Town Hall meetings was less than expected. Initially, we felt that the low attendance at events was due to inadequate marketing, but attendees informed us that the fear of being seen at a public forum on

HIV/AIDS, or being observed while being tested was more likely the cause. Another example of SFD was evidenced by the fact that many of the HIV/AIDS organizations in these rural settings do not have signs advertising their services.

All of these examples of SFD can be extremely devastating for a disease that has already approached epidemic proportions in African-American communities. Consequently, the disease will continue to spread in the African-American community unless approaches and strategies are used that encourage open and nonjudgmental discussions about risk behaviors and risk reduction behaviors, as well as increased knowledge of HIV and AIDS status. Therefore, what can be done to overcome this strong sense of fear and denial of the disease in rural settings in order to increase HIV/AIDS intervention rates?

We must first begin to understand the reasons for the strong denial associated with the disease. If we don't, we will only be spinning our wheels in the fight against HIV/AIDS in the African-American community. In fact, understanding some of the underlying issues may continue to provide a changing framework of prevention for how we plan and implement prevention activities for HIV/AIDS in the future.

One aspect of the stigma associated with HIV/AIDS probably surrounds the initial characterization of the disease in the early 1980s. Many still carry the early myths of an "immediate death sentence" with a diagnosis of HIV/AIDS. They also carry the indelible images of the initial risk groups for HIV/AIDS: gay, white men who have unprotected sex, people who share needles or paraphernalia during intravenous drug injection, or those who receive blood transfusions.

Despite recent broadening of these initial risk groups to include people of color, mainly Black and Latino, from urban areas in the Northeast to include those living in South and in rural areas and those engaging in unprotected heterosexual activity, including women, most in the community can not get past those initial images. It is as if being characterized outside these boundaries somehow protects one from contracting the disease.

In fact, the other underlying cause of this denial is that behavior is often not the focus of the community, but labels and stereotypes are. Of course, many aspects of prevention of the disease are difficult issues for most to handle: homosexuality, substance abuse, premarital sex, the sex worker/prostitution industry, extramarital affairs, and prison issues. Unfortunately, these issues are even more difficult to handle in the African-American community.

For example, acknowledgement of alternative lifestyles, such as bisexuality and homosexuality, must exist to begin to tackle some of the basic risk factors and risk reduction behaviors that are essential for prevention of this disease. A lack of openness in addressing alternative sexual orientations and behaviors probably inhibits certain preventive strategies and messages needed for prevention of HIV/AIDS in African-American communities.

Another aspect of the fear and denial that must be addressed involves dissolving the "us vs. them" aspects of persistent HIV/AIDS stereotyping, which prevents the internalization of HIV/AIDS prevention messages. During educational sessions, I often lay out all the risk behaviors for HIV and ask people to consider if they have engaged in any of these behaviors or had sex or shared needles with anyone who has engaged in these behaviors since 1981. In order to eliminate moral or judgment barriers, I often admit to my audience that during my younger years, even I have engaged in behaviors that put me at risk, but only by the grace of God have I not been affected. The audience is often shocked that I would expose myself, and the admission opens the door for true and open dialogue around very sensitive subjects.

I think more of us in all segments of the Black community need to be willing to be truthful and open about the issues that intersect with HIV/AIDS. In a recent session with young people in a Summer Youth Camp at a local church, the presenter described a scenario that is familiar in every Black community in order to make real the behavioral risks of intravenous drug use in contracting HIV/AIDS.

He described how a person addicted to crack or heroin would do anything for the drug - steal from relatives and sell stolen contents, sell their bodies, or beg for money. A few of the kids laughed. I asked, "What was so funny?" They couldn't say.

I'm not sure if it was nervous laughter because it reminded them of people they knew, or if it was because they personally did not know the persons and they were unable to relate. I suspect it was likely a little of both. The presenter and I told the group that we both had relatives who have substance abuse issues and I suspected that many more Black families did as well. Certainly other pertinent issues in HIV/AIDS were familiar to most families, including incarceration rates. However, many were not able to translate these facts, which certainly affected them on a daily basis into risks to be avoided in their own lives. It was as if the statistics were not real. And we certainly cannot rely on our elected officials. If we did, we would be in trouble.

Most are not aware of the HIV/AIDS epidemic in the African-American community, as evidenced by the recent 2004 Vice Presidential debate. Both candidates did not adequately address a direct question posed to them about the problem here in the United States. In fact, one of the candidates spoke more knowledgably about the problem in sub-Saharan Africa than here at home.

Those of us in public health are focused on getting basic health information to the masses, and that within itself is good. More efforts to get basic information about what HIV is, how it is transmitted, and how it can be prevented needs to continue. But the information by itself is not enough. Our prevention efforts have to be reframed to first understand and characterize SFD, and to begin to develop strategies to dismantle SFD in order for prevention messages to be effective.

For example, many approaches use the message, "HIV/AIDS can happen to anyone," with pictures of a diverse group of people. However, a more direct message could say, "Don't believe the myth that HIV/AIDS only affects white, gay men," or "A HIV diagnosis is not an automatic death sentence.....you can test positive and live a lengthy, successful life."

Additionally, prevention campaigns that aggressively address gender orientation and sexual behavior, as well as intolerance, would go a long way in opening up the community to first acknowledge risky behaviors and then to prevent them. Other messages that directly address issues of stigma, fear and denial may be the first steps in eliminating this disease in communities of color.

All segments of the African-American community have to get involved in order to change the frame of reference around issues of stigma, fear, and denial before prevention will work. That does not negate the fact that many more resources will be needed to eliminate this epidemic in the African-American community. The recent rural Alabama HIV/AIDS tour and current statistics emphasize the fact that the prevention resources have to be expanded in nontraditional HIV/AIDS settings, such as smaller, rural communities, and to non-traditional audiences, such as Black women. But we in the community have the greatest weapon to beat this disease in our own homes and communities by changing our own attitudes and beliefs about issues of stigma, fear and denial. I'm convinced this is the only way we will be able to prevent future cases of HIV/AIDS in the African-American community.

Discussion and Reflection Questions – Chapter Five

1. Review the facts about the United States Public Health Service Syphilis Study (a.k.a. Tuskegee Syphilis Study) and its current day implications for distrust by African-Americans for the health care system. Give or discuss any other examples where African Americans distrust the system (i.e., AIDS genocide theories – where did AIDS come from and is the creation of government to kill Black people?)

2. How can these issues of distrust by African-Americans be turned around to eliminate HIV/AIDS in our community?

3. Do you have negative stigma, fear, or denial (SFD) around issues of HIV/AIDS? If so, where do you think these feelings originated from?

4. If one has negative SFD, what strategies could be used to decrease it in you or in others?

5. Is HIV/AIDS openly discussed in your hometown? If so, what do you think

contributes to this? If not, what do you think contributes to this?

6. What can you do personally to decrease SFD in the African-American community?

Chapter Six

Healing Lessons: The Church

The Necessary Ingredient

And now abideth faith, hope, charity, these three; but the greatest of these is charity.
I Corinthians 13:13

Have you ever baked something and it just didn't turn out right? (You can raise your hand silently so you won't embarrass yourself.) I can't tell you how many times it has happened to me. Sometimes you can leave ingredients out and it will not affect the recipe that much. For example, a friend of mine, who is a baking and cooking connoisseur, worked diligently on a cake. She sifted the flour and painstakingly measured all the ingredients. She used the best of ingredients - buttermilk, flour, and butter. The cake looked perfect. The icing was scrumptious (I tasted it before it was finished). We couldn't wait to eat that cake.

I had the privilege of cutting the first slice. I couldn't wait. I cut into the cake and it was brown. it wasn't a carrot or spice or apple cake. We all gasped. What happened? We knew it was supposed to be a white cake. My friend ran to get the flour bag and discovered that instead of buying enriched bleached white flour, she had brought brown unbleached flour. Although the cake looked horrible, it tasted fine. The switch in flour did not change the quality of the cake. We teased the cook, because she worked so hard on the cake.

In my 15 years of work with AIDS, I am convinced that the Black Church plays one of the most pivotal roles in control of this disease. However, controlling AIDS will take the necessary ingredient. I believe I Corinthians 13 provides a prophetic word for the church. The necessary ingredient is love for mankind and those who may not behave or act like us.

When I conduct AIDS 101 sessions, I get lots of questions that focus on a person's curiosity about the disease and not on prevention. Much energy is focused on things like:

- Can I get HIV from kissing an infected person?
- Can I get HIV from hugging and infected person?

- Can I get HIV from the utensils of an infected person?
- Where did the virus originate?
- Do you think the virus was produced to conduct genocide against Black people?
- When will there be a vaccine?

Very rarely, does one ask:

- What are the most risky behaviors for contracting HIV/AIDS?
- Am I using a condom correctly?

- What are some of the most common mistakes made in preventing HIV/AIDS?

God always wants us to focus in on what is most important – love and not legal commandments or rules. The commandments provide structure and discipline and a roadmap for life. But it is love that dictates the behavior of the righteous. Therefore, I think that love plays three major roles and functions for the church in this struggle with HIV/AIDS.

First, if the church is the safe haven of the community as God has commanded, then every

Black church in this nation and the world should make the elimination of HIV/AIDS one of its major priorities. Secondly, the church should be a safe haven, welcoming and encouraging for those infected with AIDS, whether inside and outside our congregations. Lastly, love for those who are hurting or suffering from HIV/AIDS, should supersede everything.

Say you were the pastor of a church. Could you imagine what would happen if you came to church and had to tell your congregation about your HIV status? Would your congregation display an attitude of love and encouragement? Or would some leave saying, "How did he get it? I knew he was gay"? I would hope most would spend their efforts trying to find ways to help. If churches used the necessary ingredient to help those who are currently infected, and they could openly discuss ways to prevent others from becoming infected.

Secondly, we must focus on what is required to control this disease. For example, it is true that many of the risk factors for contraction of the disease are moral matters - homosexuality, promiscuity, drug abuse, and adultery, which are all major risk factors for HIV/AIDS.

I have conducted many HIV/AIDS sessions at churches where parents boast about talking to their kids about sex and HIV/AIDS and that child will come to me after session and say, "No they didn't". The conversation is usually very limited to "Don't do it," and often parents condemn and intimidate rather than educate their children about effective ways to protect themselves. The conversation with our children should be based on love; that is, what do those risky behaviors mean and how can we help our children understand how to protect themselves.

Lastly, the church should be the model that eliminates the extreme denial that currently exists in the Black community regarding HIV/AIDS. A couple of years ago, I was speaking to a group of Black pastors/ministers and a minister from South Africa stood up and said, "You know that the church was in denial about HIV/AIDS in South Africa until we discovered we were burying persons with AIDS on a daily basis." The myth that HIV/AIDS is a "gay, white male disease" and "can't happen to me" is exactly that - a myth. Members of congregations are dying of HIV/AIDS and parishioners are refusing to speak out about the disease, mainly because of the fear and often negative stigma associated with the disease.

The Black Church should not be the reason that those who are infected cannot be a testament to those who are living successfully with HIV. Can you imagine the power and message sent if every Black pastor in America got tested for HIV and reported their HIV status to their congregations and then challenged their congregants to also get tested and know their HIV status? So many people are walking around not knowing their status and spreading the disease. If everyone got tested, significant steps could be taken toward controlling the disease. People would be talking, getting educated and empowering our community.

The message of prevention could not be spread about HIV, but about so many other diseases such as:

- Breast Cancer - Women getting their mammogram, doing their breast self-examination and getting their clinical breast exam

- Prostate Cancer – Men getting their digital rectal exam and prostate specific antigen results
- Heart disease - Adults knowing their cholesterol levels and low density lipoprotein (LDL)/high density lipoprotein (HDL) levels to control heart disease as well as reduce their risk for diabetes
- Diabetes – Adults as well as children knowing their body fat mass content to control the current epidemic of obesity, a

major risk factor not only for diabetes, but heart disease and cancer

There is power in love. Who better to show the power in love than the church, and effectively use this power to control HIV/AIDS in the Black community? 1 Corinthians 13: 10-13 says, "but when that which is perfect is come, then that which is in part shall be done away. When I was a child, I spake as a child, I understood as a child, but when I became a man, I put away childish things, For now we see through a glass, darkly; but then face to face; now I know in part; but then shall I know even as also I am known. And now abideth faith, hope, love, these three, but the greatest of these is love."

The Desire to Be a True Servant*

Have this mind among yourselves, which is yours in Christ Jesus, who, though he was in the form of God, did not count equality with God a thing to be grasped, but emptied himself, taking the form of a servant, being born in the likeness of men. And being found in human form he humbled himself and became obedient unto death, even death on a cross.
: Philippians 2:5-8 (NIV)

***Appeared in the *Health Reporter,* the Newsletter of Frontiers International Foundation Inc., December 2006**

One of the pleasures of working at Tuskegee University (TU) in Tuskegee, Alabama is the sacred history that bestows its walls: The founding of TU by Lewis Adams, who acquired many acres of land and set the foundation for the philosophy of Booker T. Washington; the work of George Washington Carver; the important contributions of students and faculty to the Civil Rights movement, particularly through important work of the Tuskegee Civic Association and legislation that set the foundation of the Civil Rights Act; and the bravery of the Tuskegee Airmen. This glorious history sometimes gets overshadowed by the infamous Tuskegee Syphilis Experiment, renamed the U.S. Public Health Service Syphilis

Study to shift primary responsibility for the experiment from the university and town to the government.

The gross mistreatment of hundreds of Black men from Macon County in the name of science still remains a symbol of racism in medicine and causes those of us in health care to try and reverse the distrust that Blacks and other ethnic groups have of the medical community. Despite the negative legacy, the many positive legacies of Tuskegee continue to amaze me. The fact that one place has been home to so many occurrences that have advanced the plight of Black people is certainly a gift of God.

Recently, I was privileged to be a panelist on the same program as Colonel Carter, one of the original Tuskegee Airmen, and his wife, an accomplished pilot in her own right. The dignity and grace of the Carters became immediately apparent with each word that they spoke. As each began to tell their stories, I gained a deeper appreciation for the courage and sacrifice that they made in their lives to make others lives better. This became clear when Colonel Carter was asked, "How could you go and fight a war for a country that would not treat you as an equal when you came back?"

The interrogator, a relatively younger man said, "I find it hard to go over to Iraq, even today, when I know that people treat me differently than my white counterparts. I couldn't do it!" Colonel Carter said very quietly, "To tell you the truth, what we were trying to do was better the conditions for our race, so how we personally felt didn't really matter."

Scripture emphasizes this point. Jesus was the model of a true servant; he never focused on himself and he always had a bigger purpose in the eyes of God. Certainly, Colonel Carter's response is in the spirit of a true servant. He said, not me Lord, but the lives of my people. How can we apply these principles to our current crisis of HIV/AIDS in the African-American community? First, we have to recognize that there is a problem. In the 1980s, HIV/AIDS was characterized as a problem that affected small segments of the U.S. population. If you were a white gay male, you were at risk, if you had a blood transfusion – you were at risk. If you injected drugs and shared needles, you were at risk.

If your pregnant mother was infected, you were at risk.

Fast forward 25 years and the picture is slightly different now that the national blood supplies are tested, a practice that began in 1985. The risk in transmission from mother to her unborn child has dropped with increased testing of pregnant women and the use of HIV medications such as azidothymidine (AZT). The risk in white gay men has decreased; however, the risk for black men as well as women has increased dramatically, as the result of both homosexual and heterosexual transmission, and continued intravenous drug use.

The numbers are staggering. According to the Centers for Disease Control and Prevention (CDC), in 2004, African-Americans made up 50 percent of new cases of HIV/AIDS. Of all men infected, 50 percent were African-American and 70 percent were African American women. In other words, African American women have an increased risk of infection compared to white women, as do African American men compared to white men.

Despite the overwhelming numbers, I don't think that most African-Americans quite understand the magnitude of the problem. It doesn't help that our elected officials don't have solutions to the HIV/AIDS problem. For example, during one of the publicly televised Vice Presidential debates, the moderator asked candidates about the increase of HIV/AIDS in African-American women. One candidate had no clue; the other gave a long dissertation on the AIDS epidemic in Sub-Saharan Africa, but provided no insight into the problem in the United States.

Pamela H. Payne-Foster

The government is giving African-Americans limited support, and the solutions to the HIV/AIDS epidemic lie within our own community. If that's the case, what is holding us back from addressing the problem? Perhaps it is the syndrome that Colonel Carter addressed and one that Jesus talked about: "If it ain't in my backyard, then it ain't in my backyard and I don't have to deal with it."

Issues like fear, negative stigmas, and denial stem from us tackling the difficult issues attached to HIV/AIDS, such as homosexuality and the Down Low syndrome where seemingly straight men – many in the church even - cheat on their wives and girlfriends with other women or other men. The rise of AIDS is also connected to the high incarceration rate of our men and their increased risk of contracting HIV/AIDS while in prison and infecting their loved ones when released back into the community. Instead of insisting on advocating comprehensive HIV/AIDS education and treatment programs that include condom distribution, – which have been shown to decrease HIV/AIDS rates, we just ignore the problem. Or maybe we don't want to deal with intravenous drug abuse (IVDU) in our cities and neighborhoods and advocate for needle exchange programs that have been shown to decrease HIV rates. Perhaps it is easier to just ignore the problem.

The impact of the disease on our community has already begun to hurt us. Living in Alabama used to be protective for this disease. In the 1980s, the disease was seen mainly in large urban cities with populations over 500,000. However, the South and rural areas are seeing a large increase, particularly in African-American women.

"Is There a Balm in Black America?"

Hopefully by now, I have convinced you that there is a problem. So now the question is, Women of God, what are you going to do about it? Both the Colonel and Jesus gave us some clues. The answers may be in the spirit of servanthood. In other words, it's not about us as Colonel Carter would say, "It's about uplifting the race, building up the community, edifying humankind." Jesus would say, "It's all about God, his purpose, and the power of compassion and love."

People already infected with HIV/AIDS need us to be servants. They don't need condemnation, they don't need to be preached to, and they don't need negativity. They need us to serve them – to see them and let them feel our love and compassion. It reminds me of times in my life when I needed love in times of trouble. I remember when I was 16 years old and ran the car into a neighbor's house.

My mother's concern about my health and safety affected me much more strongly than my father's cold silence and obvious anger at the fact that I wrecked the car. My mother's words, "We can always get another car, we can't replace a daughter," encouraged me to overcome my fear of driving again. Don't get me wrong, I appreciated the strict discipline of my father. But at that time, my behavior, that of an inexperienced driver who panicked, needed more than discipline.

It was love that guided me through the crisis of the car accident. Jesus taught over and over in the scriptures that love and compassion save and heal. II Corinthians 12 says without love we are nothing – "as a clanging cymbal". We can't ignore the problems of people who are in need if we are true servants. If we are true servants, we will run to the problems and the people and exemplify what Jesus and Colonel Carter have modeled for us. Lord, make us desire to be a true servant in the elimination of HIV/AIDS in my people. This is my prayer. Amen.

The Prophetic Role of the Black Church in HIV/AIDS

Behold I come quickly; blessed is he that keepeth the sayings of the prophecy of this book.
Revelation 22:7 (NIV)

As mentioned earlier, a couple of years ago I was speaking to a group of Black pastors/ministers when a minister from South Africa stood up and said: " You know that the church was in denial about HIV/AIDS in South Africa until we discovered we were burying persons with AIDS on a daily basis." I hope that we as African-Americans do not wait to move into action in this current HIV/AIDS crisis. I hope that we can learn from the powerful statement of this South African minister and move to a more proactive stage when it comes to this disease.

The Bible gives us example after example of people who didn't hear God until it was too late. It was the prophet who heard God's word before the rest of the people and attempted to spread the Holy Word onto the rest of the community. Like the prophets, the church, too, must in some way hear the word of God before the rest of the community. What is the church's prophetic vision and the church's role in control of the current HIV/AIDS epidemic in the Black community?

What rhema word or prophetic voice is God trying to reveal to the church concerning HIV/AIDS? Recently, the Tuskegee University and University of Alabama EXPORT conducted a tour in Alabama's Black Belt region. We went to five cities in the Alabama Black Belt, three historically Black Colleges and Universities (HBCUs) in Alabama, and to two high schools to take the message of HIV prevention. This trip was unique and revolutionary because most HIV/AIDS prevention strategies target urban settings, not rural ones.

Even though the highest HIV/AIDS rates have traditionally been in cities, the rural population should not be led to believe that they are free of the epidemic. Of the top 10 Alabama cities with HIV/AIDS, six are in the Black Belt and in rural towns and cities. During the tour, we saw the effects of the HIV/AIDS and the negative stigma often attached to the disease. In fact, people were afraid to come out to Town Hall meetings to even get information for fear that someone would see them and connect them with the disease.

The prophetic word for the church is this: The people perish from a lack of knowledge, misinformation, myths, and fear; the people perish for a lack of knowledge. For example, if you ask a person about the ways that HIV can be transmitted: he or she would say through the exchange of blood, vaginal fluid, and semen. Most would leave out the other transmission routes of breastfeeding and through the womb to an unborn child. Some would instead say other routes include saliva and kissing, hugging, sharing utensils, and blood transfusions. They would be wrong. They are a part of the "what if crew." They spend a lot of time and energy talking about these very minuscule chances of contracting the disease, for example, asking if a person who is infected has a cut on their lip and they kiss them, can they get AIDS? And they will completely ignore high - risk behaviors such as having unprotected sex with an infected person.

The other prophetic word for the church is that it is not who we are but what we do that puts us at risk for HIV. When the disease first came on the scene in the early 1980s, the disease was characterized as an illness for gay white males. Then intravenous drug use (IVDU) became a major risk. Those who received blood transfusions, including surgery patients and hemophiliacs were eliminated by 1986 when the blood supply began being tested for the virus. Then in the 1990s, the disease shifted from gay white men to men and women of color. Although behaviors such as men having sex with men (MSM) and IVDU remain high risk factors, now heterosexual contact has become a common occurrence.

For example, a typical HIV case today might be a married 40-year old African-American woman. No matter what your status, whether you identify yourself as single, married, gay, straight, bisexual, church deacon, street drug addict, professional drug addict, politician, minister, or Sunday School teacher, a man who has unprotected sex with a man puts himself at risk of HIV. The disease doesn't care who you are - a NBA basketball player, a NFL football player, or a musician - because risky behavior puts you at risk. Many people so sincerely believe that HIV/AIDS is a gay white disease that they do not even examine their own behavior, because if they are not white, they feel they are immune from the disease.

So obtaining all the facts, sifting through and disregarding misinformation, myths, stereotypes, and prejudices and fears about HIV/AIDS will move the church to a new level in what God promises: the healing and wellness of his people and the people of the land. The church needs to have an ear to hear what the Lord is saying in regards to this epidemic and the healing of his people. Amen!

Pamela H. Payne-Foster

The Important Healing Role of Spirituality in HIV/AIDS

Is there no balm in Gilead? Is there no physician there? Why then is there no healing for the wound of my people?
Jeremiah 8:22 (NIV)

I first started working in the HIV/AIDS field while training as a medical resident at the State University of New York at Stony Brook (SUNYSB) in Long Island, New York, in 1990. The disease was still relatively new and still being characterized. Although New York City was considered a epicenter of the disease, many infected persons who were originally from the Long Island area, a suburb of New York City, came back home to die.

Therefore, treatment, hospice care, and other policy issues, such as confidentiality, particularly in discrimination of jobs, schools and financial issues such as insurance and health care, were being hotly debated and played out by advocates, mainly, the white gay community. Because the volume of AIDS patients was large, SUNYSB actually had an AIDS unit for long-term and pre-hospice care.

Pamela H. Payne-Foster

I vividly remember the rigid isolation procedures and the sense of hopelessness at the hospital in which people came to die. In 1991, I was assigned to work with the first Director of HIV Bureau in Nassau County, N.Y., Dr. Brian Harper. So many of these issues were being played out during my training. I saw first - hand many of the policy issues that unfortunately still remain today, fueled by strong negative stigma and fear attached to the disease.

For example, I spent many hours talking and providing AIDS 101 sessions to PTA groups, school administrators, teachers, and school staff to alleviate their fears about children with HIV attending schools in their districts. I also remember calling Black churches to offer HIV/AIDS 101 educational sessions to their ministers and congregants and running into occasional resistance on all or certain parts of the prevention messages. For example, they would allow a session if I didn't mention preventive strategies such as "condoms".

After my residency training, I relocated back to the South and joined a church. At the time that I joined in 1993, the pastor asked me to be the first

Health Ministry Chairperson and head the Health Ministry. I was excited. I quickly assembled a ministry committee, a group of experienced health professionals and those interested in health, which included several physicians, lots of nurses, and dozens of federal health employees.

The advisory board included very high - profile health and public safety officials. The pastor suggested that we conduct a survey of the congregation. The survey was short and open-ended. We asked the congregants to write the health priorities that they would like to have the church address and, surprisingly, HIV/AIDS was the Number 1 health concern of the congregation.

The health ministry committee included one elder with the rest of the committee made up mainly of laypersons who were not in church leadership positions. One of the challenges of our committee was to encourage persons with secular knowledge to translate that knowledge into a spiritual mission. The success of this bridge or transition relies on established church leadership that may not be comfortable with health information that is packaged in very secular messages. Therefore, it was invaluable to have an elder on our committee. The elder took the intentions of the committee to the church leadership and took issues from church leadership to the committee.

One example of the challenge of merging heath education/prevention messages with the spiritual mission of the church is that the pulpit in the Black Church is a position of power. Because of this, my committee felt very strongly that the pastor should preach a HIV/AIDS prevention message. My pastor agreed, but was very uncomfortable delving into a topic that was not in his area of expertise. The pastor also felt some reluctance, especially in the area of prevention: How, he asked, could he preach about the use of condoms when every other Sunday he preached abstinence. In the same way, the pastor would allow us to pass out condoms at a health fair, but wasn't completely comfortable with it because he felt as if we were promoting sex.

I am not a pastor, but as a child of God who has worked extensively in the African American community around the HIV/AIDS issue for more than twenty years, I am frustrated that the epidemic is worsening in our communities. What advice can I give to make a difference? Some thoughts: First, no one else can fix this problem, but us; no one else cares. The government doesn't know or care about it (blatantly obvious during Vice Presidential debate).

The Black community has to think of ways to make the knowledge and prevention of HIV/AIDS a major priority and translate that knowledge into decreased risk or harm reduction strategies. Those of us in prevention need the community to assist us in developing health education campaigns aimed at preventing the disease, provide increased access to testing, and provide better access to treatment so that those infected can live longer and better quality lives.

While the traditional prevention techniques are good in theory, they haven't worked in our community and I don't think they are going to work. I offer two other approaches to assist in this epidemic in our community. First, until we remove the negative stigma attached to the disease, and the negative stigma attached to the risk factors, we can forget about preventing HIV/AIDS. The climate should be such that people can talk about the disease very casually with little worry about what people think.

That is not the case even now. When we tested students at Tuskegee University, those waiting to be tested were harassed by fellow students (many of them football players) with lines like, "What you getting tested for, you got AIDS?" Knowing your AIDS status should be as natural as knowing your cholesterol level. It is just a matter of health.

Similarly, when talking to audiences about risky behaviors such as anal sex, men having sex with men, or even oral sex, people are very uncomfortable. Sexuality has to become easier to talk about, especially in the church. Even as a seasoned HIV/AIDS educator, it blows my mind to have to talk to fifth graders, some of whom are already sexually active, very graphically about HIV/AIDS prevention because that is how young children are engaging in sex now. I suspect that some of these children may be in Sunday school; some may not. So our message has to be everywhere we are: church, beauty shops, work, school, etc.

Secondly, the message of prevention has to include why screening for any disease is important, including HIV/AIDS. The "I don't want to know" syndrome is prevalent in our community. Why should an African-American male get screened for prostate cancer? A common feeling is, "If I got it, I don't want to know." There are many examples of African-Americans having a lower quality of life and dying earlier because they have diseases and don't get early screening to prevent complications and premature death. For example, if African-American men would get an annual prostate specific antigen and digital rectal examination after age 40, many could be diagnosed early, receive treatment, and therefore, add many potential years to their lives. The blessings that come with that fact should outweigh the fear of not getting tested. My father is a living witness to that fact. He is healthy nine years after his diagnosis of prostate cancer, unlike at least a half dozen of his friends who died an early death.

The examples of people who have been diagnosed with HIV, who as a result of early testing and diagnosis are still living successfully with HIV, are plentiful. Magic Johnson is one of the most famous. I had the opportunity to hear Magic Johnson speak in Birmingham in 2004, and although everyone knows he is rich and, therefore has access to the best medical system available, he attributes his longer life to his positive attitude and increased spiritual maturity. This is a powerful message to our churches that will influence prevention and also act as a safe haven for those who are infected.

Even in 2007 I still hear horror stories about churches and families shunning those infected with the disease. Church leadership and health professionals working together to improve the health of their congregations and the communities they serve should be the ultimate goal in AIDS prevention. There is an important role for the church and its foundation of spirituality in the healing of the community suffering from HIV/AIDS.

Discussion and Reflection Questions – Chapter Six

1. What role do you feel the church plays or could play in healing and wellness, particularly as it pertains to HIV/AIDS?

2. What is your congregation or others in the community doing to address HIV/AIDS, particularly in the African-American community?
3. Do you feel that the Black Church should play a prominent role in the elimination of HIV/AIDS in the Black community?
4. What do you feel are some of the barriers that the Black Church would have to overcome in order to eliminate HIV/AIDS in the African-American community?
5. What are some strategies that could be used to encourage and support the Black

Church in the elimination of HIV/AIDS in the

African-American community?

Chapter Seven

Healing Lessons: The Family

The Unveiling of "Family Secrets" in Health: The First Step in the Healing of the Black Family

Is any one of you in trouble? He should pray. Is anyone happy? Let him sing songs of praise. Is any one of you sick? He should call the elders of the church to pray over him and anoint him with oil in the name of the Lord. And the prayer offered in faith will make the sick person well; the Lord will raise him up. If he has sinned, he will be forgiven.

Therefore confess your sins to each other and pray for each other so that you may be healed.

Pamela H. Payne-Foster

The prayer of a righteous man is powerful and effective. James 5:13-16 (NIV)

If your family is anything like mine, it has either one big family secret or a couple of small ones. All you have to do is remember back when you were a child and you heard the grown-ups talking, but they were really whispering, and sometimes they might have been talking in grown-up codes, like spelling things. But if you were like me, you would try to figure out the spelling or might even be so bold as to ask, "Mama what does "h-u-s-s-y" mean?"

The more experience I get in my profession of healing and medicine and in my spiritual walk, the more I realize that the area of hiding illnesses and diseases is a major dysfunction in the African-American community's ability to deal with certain lifestyle behaviors or conditions that affect our health. This is not unique to us as African-Americans.

The same shame that prevents diseases like cancer and HIV/AIDS, for example, from being opening discussed, prevented or treated affects white families, too. But we as African-Americans die more often than whites from HIV/AIDS. These vast disparities dictate how HIV/AIDS get played out in our communities, making this topic much more crucial for us to discuss.

I very strongly believe that one of the first steps in healing is acknowledgement. I believe you've got to verbally acknowledge disease in order to be healed. In fact, from a spiritual context, the power of prayer is to verbalize it. Healing is less effective if you think it or meditate on it in silence. It is in the crying out that God can hear it.

The negative stigma attached to HIV/AIDS is still prevalent today and, unfortunately, I still hear people say, "If I have a lump in my breast or have some symptoms I don't want to know that I am sick or have cancer." This attitude still prevails, even though most of us know that the earlier we get diagnosed, the better our chances of survival. Perhaps we need to spend more time helping people understand the power of early and secondary prevention.

I want to give two examples in my own family that occur in most American-American families, which I think can help heal many African-American families. I stand here jubilant that my father is a thirteen-year prostate cancer survivor. I remember when my father was diagnosed in 1994, he specifically told me not to tell anyone. At that time, I was studying the Bible and because I loved my dad, I listened to what I learned in Bible study and told me, contrary to what he asked. Scripture says clearly knock and the doors will be open, ask and ye shall find; "Ask the pastors, ministers, or elders of the church, or the spiritually mature, to pray when in need.

So when the time came for my father's surgery, I lifted up his name and his condition in a prayer meeting and God heard my prayer and healed my father. My dad didn't stay mad at me for long. In fact, after his surgery and after outliving many friends who died of prostate cancer, my father eventually went on to volunteer for the American Cancer Society and had the nerve to say to me, "Ain't no Black men up here volunteering," I laughed to myself: "No, cause they are all like you used to be, don't want nobody to know".

In the same way, the secrecy, shame, and denial associated with HIV/AIDS is killing our community. In Alabama, 70 percent of new cases of HIV/AIDS are African American community. More than 70 percent of new cases of HIV in African American women. It has passed the point of what we in public health define as epidemic. It is a crisis! But issues such as homosexuality, high substance abuse rates, and high incarceration rates in our communities are driving what was once characterized as a gay white male disease as a disease of color. Those three risk factors - men having sex with men, intravenous drug use (IVDU), and increased rates in prisons - have to begin to be addressed openly and honestly in the community before we can deal with testing, prevention, and adequate treatment.

For example, although I have a brother who is incarcerated for his long-standing drug addiction, his addiction still is not openly discussed in my broader family. If someone asks me how my brother is doing, as a healer I feel very comfortable saying,

"You know, he's in jail, but to tell you the truth, it is a blessing in disguise, because he is getting the drug treatment he needs and deserves." My letters from him give me the most hope because I know he is on the road to sustained recovery. Despite my ease, other family members find it hard to be that revealing. However, I believe that it is only in his transparent and broken state that he will be healed. To promote this healing, my family needs to be more open about his incarceration as the secrecy could quench and choke out his spirit.

Going beyond our own personal experiences, I want to encourage action to address the large issues, such as HIV/AIDS, and substance abuse from a community perspective because I know as a community, both issues will affect each of us more and more. Rockefeller laws, which imprison many of our community unnecessarily, are a travesty and we should be rising up clearly and loudly against them.

Alabama has historically had long waiting lists for people who test positive for HIV to get medications if they don't have the money to pay for them. The waiting list was eliminated with recertifications and the implementation of Medicare D in October 2006. As of December 2006, the Alabama Department of Public Health had reached full capacity for free HIV/AIDS medications.

The capacity for lack of access to treatment for some who are HIV positive still looms near and is a travesty that we should be speaking loudly and clearly against. Finally, we should be advocating for better prevention programs both in substance abuse and HIV/AIDS to make our communities healthier.

Kaleidoscopes

Then you will know the truth, and the truth will set you free. John 8:32 (NIV)

This section is a composite of fictional characters made up by the author based on real-life experiences as a public health physician and HIV/AIDS health educator, particularly during her time in rural areas of Alabama.

Emma

I contracted HIV/AIDS when I was 75 - years - old. A lot of people cannot believe it when I tell them. I guess I could have kept it a secret, but at my age, life is truly short and I want people to know, AIDS is no respecter of age or gender. If you slip up, it may be waiting for you. I remember a nice young man (about 50 years old) - ya know age is relative (laughs) – he came down from the Health Department to educate me about this disease. I didn't know nothing about it. But I probably got it from a friend named Bobby – a young man in his 30s who would come over to talk to me and keep me company and every now in then-service my physical needs – if you know what I mean. Even old ladies get lonely now and then. Well, he services lots of people, female and male, and probably didn't know he had HIV, and maybe he did, but the rest is history.

The hardest part of having the disease is not so much the symptoms but the shame that it has placed on my immediate and church family. But shoot, I'm the one with the disease not them. Let me worry about the shame. I don' already gave it up to the Lord and he has forgiven me, so why shouldn't they. I just wanted to get the word out, especially to the elders in the nursing home where I live, to watch out; HIV can infect even you. I pray to God that it don't, but it could.

Phillip

Pamela H. Payne-Foster

I am infected with HIV and nobody knows it. You see, I live in a small country town in Alabama and I don't think people can handle it. I've been to visit relatives in large cities where being gay is a little more accepted but it is definitely not like that where I'm from. The environment is downright hostile for gays. Of course, with it being the Bible Belt, I have heard every nasty cliché there is to hear about "God didn't make no Adam and Steve" and "homosexuality is not just a sin, but an abomination before God." If I hear those tired clichés one more time, I think I will vomit.

I know for a fact there are plenty of gays in the church, at all levels: from the ministers, to the deacons, to the choir members and musicians to the married men, and to the youth. I know because I am part of a secret society of those of us who know. There are plenty of ways to subdue and hide it: through status, the higher you go up in status in the community, the less likely people are to suspect – I know bank presidents, I know pastors of churches, I know college administrators, I know politicians – yes, their status protects them, but so does their marital or dating status. Most have wives or girlfriends. That's the biggest lie there is. Someone will say,

"They can't be gay, he has a girlfriend, he's married."

Right!

I even belong to an underground society of people who work in the HIV/AIDS field. Of course, I know that being open with my HIV status could be used as a powerful tool to effect change in the community, but I can't chance it. I refuse to give up on my relationships and if I revealed that I had HIV then people would talk and wonder about my sexual preferences and I just don't want to deal with that.

I really admire Black men who are open about their sexuality, although I don't know too many in this part of the country. One day maybe I will reveal my true self. I know friends who won't even do it at their funeral. I guess even then I don't want to hear people's mouths talking negatively. I can just hear my minister preaching my eulogy, "He was a great man, I mean woman, and I don't know what he was."

Cecilia

I am 20 - years - old, a college junior majoring in Biology and am living with HIV. I was born with HIV and got it from my mother. She contracted the disease as an intravenous drug abuser for many years while living in New York. My mother died in prison with the disease about 10 years ago. It's been hard not living with mother, but knowing the pain and agony that she went through, both with the disease and living in prison, has not been easy either. I have been living with my grandmother in Alabama ever since I can remember. Grandma and I have had quite a battle with this disease and I don't think the battle is over.

My physical, emotional, and spiritual health is good. You see, my grandma has covered my life. She is an Evangelist in the Pentecostal Church and she has prayed many a day over my life to let me live successfully with this disease. But it wasn't easy. And I tell you the worst battle has been with the church people. There was many a day when people wouldn't want to touch me or look at me for fear that they would get infected. It's no fun when the people of God who preach love won't give you any.

We never kept my HIV status a secret. Grandma doesn't believe in secrets. She told me many a day,

Pamela H. Payne-Foster

"If you can't verbally pray your prayer, God can't hear it and can not heal you of what ails you. So we are going to put this illness on the altar and let God have his way." So I have become the poster child and a face for young people who are Black in this country and have HIV/AIDS.

My activism started early in grade school and have been I educating not only my peers, but their parents and families ever since I can remember. You see, I had to educate myself in order to stay alive and in the process I learned that I had to be a vehicle for others to become educated and enlightened.

It hasn't been easy, but I can say over the years it is slowly getting better. But much more needs to be done. There is still a lot of ignorance attached to this disease. So it doesn't really surprise me that over the course of the epidemic, the numbers are still increasing.

Some of the other hard times for me continue to be in having friends, both male and female, who are mature enough to handle the news and mature enough to go beyond the news into relationship. But I tell you, I really value the friendships that I have because I know they are truly sincere.

Pamela H. Payne-Foster

Guest piece – Representative Laura Hall

Taken from "**25 Years of AIDS and Black America"** Black AIDS Institute – June 2006, In the Capitol: Voices from the Political Arena: Representative Laura Hall-Alabama General Assembly, Montgomery, Alabama. As told to Kai Wright

Pamela H. Payne-Foster

I was elected in 1993, and I'm 62 - years - old now. But I'm planning on being around in the General Assembly for a while. I really do enjoy it and want to believe that I make a difference. My son was diagnosed with HIV in 1988. My husband and I suspected something because he could never get rid of his cough. Eventually, we got him to go to the doctor one day. He was 22 - years – old at the time; he died at age 25. We got the results the Monday after the Sunday we had buried my dad, and my son was there visiting. We tried to talk about it, but he just said, "I don't care what you do. I don't want to talk about it anymore. As a matter of fact, I'm not even going home with you guys. I'm going back to Atlanta." He went through some difficult times after that. He started using drugs, and doing so many things out of character for himself.

We kept it secret for a long time. But when we finally told everyone, it was heartwarming. My uncle's statement was, "You know, we're family, and families stick together. Families just don't leave each other." So it was heart-warming to have that response. But meanwhile, in his own family, his son was dealing with the same disease. And his sister, she didn't know her son was dealing with the same disease. So over a period of three years, we had three deaths, and none of us had talked about it. I'm sad to say not enough has changed today, because I am very suspicious of another little cousin's condition.

By the time of my son's death, though, he had come to insist on openness about his infection. He insisted on an open casket and he made a tape that he asked us to play. On that tape, he asked everyone who was there to put whatever they had in their pockets in a basket because a collection was being made to give to an AIDS agency. He really taught us how to die with dignity.

There's a lot of work still to be done in our Black community here in Alabama. We're more sensitive to the fact of AIDS existence, and we're willing to discuss it. But there's nobody out there saying, "Oh, we need to rally around this." There's still that fear factor- the fear of being ostracized. It's difficult. I lived through that fear, so I guess I want to think that 13 years later we wouldn't have to deal with that.

But it is so overbearing. People who are where we were in our family are always worried about what people are going to think. If they are going to be there for you. I have a god-child who is HIV-positive. I know only because her mother has told me. She's 25. Now we've been in several settings where I have gone in thinking she would say something to me. Nope. She has not shared; not one time.

So I make sure that I deal with this issue in my role as an elected official. I talk about it wherever I can, and I try to bring my colleagues together to pass legislation that combats this epidemic. In my last election, Republicans even charged that all I do is talk about AIDS. I said I make no apology for my position. And if you choose not to elect me because I'm very outspoken and upfront about this disease that has impacted my family, then so be it. That means I'll spend 100 percent of my time working on this issue. Never heard another word.

Guest Piece - Heal Thyself

By Roland Barksdale-Hall

How can a compassionate heart for service transform? Select journal excerpts, written during my year-long sojourn as an AIDS Educator, tell the story of my struggle for answers to these questions. A personal story of spiritual transformation on the journey, which took me out of my comfort zone, is told in hopes we all might more closely examine those issues that keep us from significant social engagement, thereby attaining heightened levels of introspection. These journal entries explore what a service leadership track should look like. Throughout the pages, personal development in thought crystallizes. Transparent entries mark major stops on freedom roads.

*   *   *

11:30 p.m. September 7, 1999

HIV/AIDS is wreaking havoc... 1998 shows that we as African Americans are leading in the number of new cases of AIDS... AIDS rapidly is spreading through Africa. Women, men, and children have contracted he disease... We must practice safe sex, including condoms.

Recently, I applied for a temporary position as a HIV/AIDS educator. I made it through the first round of interviewing and will be having a second interview this upcoming Thursday. This assignment really would be in keeping with the spirit of ***Healing Is the Children's Bread***...I had wanted to have stories about African Americans being incarcerated and the life issues of someone confronting HIV+/AIDS in the book. The absence [clearly] was shortsighted. But, I worked with what was available...

*　　　*　　　*

10:50 p.m. Sunday, September 26, 1999

Today, I shared with Buke about [being] an AIDS educator.

--It's of God!

This is a spiritual marker.

*　　　　*　　　　*

October 6, 1999

-Friend: -"Oh you got the position at Clarion. You're going to be traveling around doing recruiting?"

-Me: -"No AIDS education."

A pause then followed.

--Friend: "Can I get back with you a little later, Barksdale."

At George Junior there were several hundred students—all male.

My locks, or maybe it was my bright, gold-trimmed vest, perhaps even the black trench coat, sparked considerable attention.

--Who is he? What is he doing here?

Mrs. C's, 9-12 grades, Health Class, George Junior Republic

"Is There a Balm in Black America?"

Define the problem

Explore the alternative

Consider the consequences

Identify values

Decide and act

Evaluate results

Do you do the grownup? [I learned that meant in today's youth culture.] Are you having sexual intercourse?

Students were frightened by the facts about AIDS.

It saddens me to think about Africa and the spread of AIDS.

*   *   *

This past week I met Andrea... She stopped by the staff lounge and greeted me in Spanish. Since growing locks many people find it hard to figure out my ethnic origin. Perhaps my hyphenated name also led to her... conclusion.

Lori, my office mate, is quite frank. She shared with me that some students at George Junior thought I was going to be the HIV+ guest.

*　　　　　*　　　　　*

5:30 a.m. Monday, November 1, 1999

These days, I find myself out of my comfort zone. The unfamiliar has become my daily companion. "What will people think?" still runs across my mind. Somehow, it no longer has the same stronghold it once did. Still, the separation pains, desire to be understood, and accepted, all along is there. No matter, the incessant desire to please others is fading, giving way to solitude.

Pamela H. Payne-Foster

There is a place in the midst of the mundane where the voice of God remains to be heard, only for those who choose to listen... Yes, I hear the voice... [T]he struggle remains to hearken... Some days it is easier than others. I'm still learning.

* * *

6:30 a.m. Monday, November 1, 1999

"I sing because I'm happy. I sing because I'm free." An exuberant song burst through the house.

* * *

5:25 a.m. November 2, 1999

Human sexuality. What is it? The PFLAG Conference posed a similar question.

The personal perspective of Christopher—he was tall and thin with the side of his head cropped; the top crop was dyed blond and trimmed in brown—touched my soul. He wore an earring and spoke in a thick Southern accent. His story was about a search for identity.

There were several tragic episodes. His youth was spent being taunted as "sissy," later "faggot." His slight build and blue eyes and blond hair and fancy cut outfits brought out these comments.

Some likely based in jealousy. Then later came the pats on the behind in the hallway and the violation. He was raped at age 12 by a man, bringing on more questions about his identity. When he decided he was gay, his father, a staunch Southern Baptist, disowned him and threw him out the house. This led to meeting gay friends in chat rooms and more victimization.

His personal perspective struck a deep chord. The insensitivity of the church to those searching for an identity—even [through what maybe perceived as] unorthodox means—[has been a sad commentary.] The church has appeared [in] the dark. Only recently has Jerry Farwell and Dick Hatch [radio talk show host] apologized for the venom spewed at homosexuals. Perhaps, religious zealots overlooked the humanness and divine in all of God's children. And the brokenness, we all share.

Gender roles in the good ole U.S.A. further exacerbate matters. While in Ghana men walk holding hands and are intimate—not sexual—friends, America offers little hope for men to develop such relationships without being labeled. Still, there is the longing buried deep within for intimacy. Some can't ignore it.

Kinsey said there is a spectrum between male and female where we all live. Maybe, it is not just sexual but should be expanded to redefining gender identities and roles.

"Is There a Balm in Black America?"

Grandpa Jesse taunted.

--That one is just like a girl

It stuck in his crawl that I was tall and thin...

Boys like to look up to their dads [and granddads as well. However, my granddad and I] detested one another.

*   *   *

8:15 a.m. November 15, 1999

On the drive from Sharon, PA to Clarion, PA

Oh Lord,

The people suffer. Where are the priests? Is there no balm in Gilead? Only fear of AIDS, fear of shadows. Fleeting is your touch. We are but empty tombs

Blood tainted. Gone all hope of... a song, sweet caress. We are the living dead.

* * *

8:00 p.m. Friday, November 26, 1999

I value talking about real life issues, grappling with questions that don't always have easy answers. Experience has taught... everyone is not prepared to handle differences. To be honest I struggled with differences, too. ...earlier if [only] I was not so narrow-minded. Fortunately I have changed...

It is wonderful to be free. Truly, being free has made the world change around me. Relationships have been renegotiated... Other questions I have are how do we handle differences of lifestyle, religion, etc? The answer seems obvious—with respect. But I still feel as though it must be asked...

*　　　　*　　　　*

3:35 a.m. December 9, 1999

For over a week I have been meditating on Luke 14:25-33, the parable of discipleship. Counting up the cost rings true to me.

As time passes, I realize a maturation process exists. Today, I am focusing on the process. Guess I realize there is a mission for me.

Deep in my spirit I feel a call to Africa. I cannot ignore it nor do I want to. Yet, I must focus my being on what it is. I do not understand on one level. But on another level I completely do. God has a work for me. It was planted in me. Now the seed is stirring.

I will seek... If recent turn-of-events bear a clue, I will complete the leadership program at Duquesne. The majority of the work is done. All I have to do is write my thesis on The Education and Prevention of HIV/AIDS in Sub-Saharan Africa, in particular Africa. Knowing God, this will lead to the next revelation.

I feel the stirring...

*   *   *

February 22, 2000

Today, I met Apollo, a tall strong black man with a rippling voice that would command a fleet of ships. The Olympian towered over most men and was striking. Outside the Victorian Athenaeum Hotel at the Chautauqua Institution our paths crossed. With the backdrop of the pristine Lake Chautauqua, I strolled down the path; he up, holding a tiny package [and asked.]

--Can I take your picture for my travel log?

--Sure.

"Is There a Balm in Black America?"

Dinnertime Apollo showed himself to be bright and articulate. He was chosen from among 30 peers to address the issues of men of color and HIV/AIDS. The presentation was forceful, making an impression on me—and I was not alone, as judged by comments of others. [T]he presentation was simple. Yet, there was no doubt he was in total control of the situation. He stood as a testament to Sterling Brown—"Strong Black men keep a comin'."

After dinner, I promise Patty and Rebecca to join them at the "Sexier Sex Party." It was a blast.

Apollo appeared there. I was quick to commend him on his powerful delivery. He asked if I was planning to attend the dessert reception.

We headed out together. He lit a cigarette outside. It was a crisp fall evening. He began sharing about himself. It was intriguing and I intently listed. Then a pause followed.

--Are you married?

--Yes.

"Is There a Balm in Black America?"

We sat by the fireside as Apollo's story unfolded. Seven years in a monogamous relationship before.

*   *   *

March 9, 2000

The oppression of Afrikan-conscious people persists.

On my way home from the AIDS Medical Conference, which was a success, a State Trooper pulled me over. I was driving a 79 Caprice Classic and fit the profile..., except I was dressed in a red, white and black striped long sleeve shirt, trimly fitted and adorned by a plat black tie and tweed black-and-white vest.

I asked.

--What is the problem?

--Let me see your driver's license and vehicle registration.

I had them in my hand.

The trooper scolded.

"Is There a Balm in Black America?"

--You were going 73 miles an hour in a 65 zone.

I doubted the [validity of what the trooper was saying] given that I was in the no passing lane. Still, I cooperated.

--Get going, he [might as well] have said.

Instead he said.

--I'll just give you a warning. Watch your speed.

I blurted out [something] about stopping African-American [males].

--I couldn't hear you, the trooper said.

My lips ever pursed. I escaped to breathe a poem.

Mint Americana

Raccoon faced engines draw Americana;

Inside separate but equal etch slumber:

Black face porters mime errand boys;

Belly-heavy conductors haunt corridors.

At another time, the hmmmmmmm...

Of the Dixie Flyer carried;

Coffee people, shoebox lunches;

Chickens (both alive and fried);

Miles mo' on elastic bumble seats

Packed in Jim Crow cars, bushes...

Blot fo' Birmingham gals' sleep.

Movin' toilets, ain't easy, you know.

*　　　　　*　　　　　*

I drove back off on freedom roads.

Discussion and Reflection Questions – Chapter 7

1. Is anyone in your family infected with HIV and has the family had open discussion about it?
2. Do all members of your family know their HIV status?
3. Has your immediate family had a discussion about the HIV/AIDS epidemic in the African-American community and how to prevent all family members from becoming infected?

4. Have you had a discussion about the issues surrounding HIV/AIDS in the African-American community at an extended family event (i.e., family reunion, family celebration, etc.)?
5. Has your family reached out to other families within your community such as schools, church, civic organizations to discuss and education about HIV/AIDS?
6. What strategies could African-American families use in combating HIV/AIDS in the Black community?

Pamela H. Payne-Foster

Epilogue

"There is a balm in Gilead
To make the wounded whole
There is a balm in Gilead
To heal the sin-sick soul"
Lyrics from a Negro spiritual

Around the end of the completion of this manuscript, the Centers for Disease Control made a recommendation for universal testing for HIV/AIDS. I think this moves HIV/AIDS prevention to a new level, since we in health promotion know that many people are walking around infected with HIV and don't even know it. Despite this announcement, I still think we have a long way to go to eliminate this disease in the African-American community.

In researching for this book, I realized there is a paucity of organizations focused on HIV/AIDS prevention in the African-American community. We either need more organizations, or we need to mobilize existing African-American institutions and organizations, such as church groups, schools at all age levels, student at all levels, particularly high school and college age youth, sororities and fraternities and other African-American social and civic organizations. That means we need a grassroots movement or approach to solving this problem.

It is also apparent that there are few foundations or organizations focused on fund-raising specifically in the area of HIV/AIDS, prevention for African-Americans. Again, tapping into existing African-American institutions and organizations will be key. Because I have intertwined health with spiritual themes, I certainly think that the Black Church should play a pivotal and leading role in this epidemic.

Pamela H. Payne-Foster

Many African-Americans are believers of faith. We have the testimony of our ancestors in of the power of faith in bringing us, as a people, through the perils of slavery. Despite this history, many of us are not linked to the healing powers of our faith. I believe that traditional medicine has its place in physical healing, but it is the spirit that dominates over the physical and its role in healing has been underrated. But in order to heal some of the illnesses of the land, such as HIV/AIDS, we are going to have to rely on **all** that God has for us! There is a balm in Gilead and it is time for us to partake of it!

APPENDIX A

AIDS 101 and Prevention Strategies

The virus that causes AIDS is the Human ***Immunodeficiency virus (HIV)***. ***AIDS*** stands for ***auto immunodeficiency syndrome,*** which describes a variety of symptoms caused by immune dysfunction. Symptoms are varied and can affect many parts of the body including the brain, eyes, the neuromuscular system, the cardiovascular system, and the gastrointestinal system. The name of the virus describes the disease and how it is transmitted. The h- stands for human – this disease is transmitted from human to human. Because the virus is blood-borne it is primarily transmitted four ways:

1. ***Sexual contact*** – vaginal and anal predominately, or in some cases oral; heterosexual; predominately male and female, or male and male. ***Protection – abstinence (100% effective); and correct use of latex condoms (oil - based lubricants decrease effectiveness) (95-99% effective).**

***Note – It is believed that co-infection with other STDs, such as Chlamydia, syphilis or herpes increases the transmission of HIV. Therefore, a prevention strategy for all STDs is important for HIV/AIDS prevention.**

2. ***Intravenous drug use (IVDU) or "shooting up"*** – sharing unclean needles

   ***Protection – needle exchange programs and use of sterile needles, use of bleach in cleaning of needles**

3. ***Prenatal transmission*** – Transmission from mother to child during pregnancy

***Protection – Use of antiretroviral drugs can decrease chances of transmission from 70 percent to 20 percent.)**

4. ***Breastfeeding*** – Transmission from mother to child

***Protection – women should refrain from breastfeeding if HIV+**

(Note: HIV used to be transmitted by blood transfusion, but the risk is nearly negative now in United States since the testing of the blood supply since 1986.)

Common myths for transmission

x Kissing, eating utensils, sweating (Although HIV is found in small quantities in certain body fluids, it is not in sufficient quantities to cause infection)

x The transmission from infected health care workers (HCW) to patients or from patients to HCW is common (There has only been one reported case of a dentist transmitting the virus to patients, so this route of transmission is rare and not well documented; the risk of transmission in HCW is also relatively low and is further decreased with the use of ***universal***

precautions – protective barriers such as gloves, goggles, health gowns in dealing with infected persons.

- The transmission of infected persons to household members is high (the risk is low and is further decreased with the use of universal precautions, which should be used by noninfected persons, and care taken in sharing of toothbrushes or personal items such as shavers, or touching sores or rashes, which may have tinges of blood, should be avoided. Additionally, sharing utensils that have been used in piercings or tattoos should also be avoided.

The I-in HIV stands for immunodeficiency which describes what the virus does to the body to cause disease. The virus works to cause the immune functions of the body to become dysfunctional. The immune system is important in protecting humans from certain fungus, cancers, bacteria, and other foreign objects. When it does not function, we are susceptible to illnesses that can be harmful. Essentially, the HIV virus attacks a subpopulation of immune cells called, CD4. That is why the CD4 count and the viral load are used to monitor an infected person. You want the viral load to be low and the CD4 count to be high in order to have fewer symptoms. Additionally, attempts to boost the immune system such as good diet, exercise, and consistent medical support, are important.

The V-in HIV stands for virus. The virus that causes AIDS was discovered in the late 1980's. After the virus was isolated, the structure of the virus was used to develop testing tools for better diagnosis. Tests are available to test the virus directly, but most of the current tests measure the virus indirectly by measuring antibodies to the virus, which the body makes after infection. Because it is an indirect test, it takes about six-eight weeks for antibodies to be formed, so there is a **window period** before antibodies appear in a test. Testing modalities have ranged from blood to saliva and testing time has been drastically decreased from days to 20 minutes (rapid testing). Viruses are tricky. They change and adapt to their environment. That is why it has been difficult to formulate a vaccine. The other challenge for vaccine development is that there are different forms of HIV throughout the world (i.e., HIV-1 is predominant in the United States, while HIV-2 is predominant is Sub-Saharan Africa). Despite the difficulties, attempts to formulate a vaccine for HIV prophylaxis or prevention are underway.

Currently the only cure for the disease is prevention. The first step in prevention is to know you and your partner's HIV status. The only way to know is to get tested. The amount of testing is dependent on your risk. After testing, I urge you to review the **bolded strategies for prevention listed above** if you engage in any of the transmission routes, and remember that the strategies can only be effective if used correctly. For example, in my experiences, many persons never learn how to use a condom correctly or consistently, therefore prevention is compromised.

Current pharmaceutical treatment for HIV/AIDS includes various types of antiretroviral medications, which prevent viral replication at various levels. Consistent use of medicines along with good overall health can prolong the quality of life and length of life as evidenced by the millions of persons who currently live with the virus.

Current statistics of HIV/AIDS in African-Americans can be seen in Figures 1 & 2. (Centers for Disease Control and Prevention). There are many theories for the gradual increase in the disease in African-Americans including: increased sexually transmitted diseases in African-Americans and link with HIV, poverty which decreases access to primary health care and prevention, the "Down Low syndrome" where men engaging in homosexual behaviors do not. inform their heterosexual partners, high risky behavior in young people in their teens and twenties who experiment with sex and drugs combined with ignorance and use of proper and consistent use harm reduction behaviors such as abstinence, condom use, sterilization of drug works, needle exchange programs, etc. Other cultural issues such as distrust of medical establishment and lack of HIV testing, negotiation of condom use by women, machismo attitudes among men about condom use, and other cultural factors may also play a role in affecting the Black communities HIV rates.

The information for this section or additional information can be found on the Centers for Disease Control and Prevention website: www.cdc.gov.

APPENDIX B

Articles, Guidelines, Sample Activities, Other Resources

The little town that could

By Mary-Ellen Phelps – Christianity and Crisis, July 4, 1988 – out of print"

In the hot, dry summer of 1987, a small Maryland town of 2300 residents said no to fear and ignorance. Instead, Denton faced the truth about AIDS, refused to dignify the disease with panic and, in its own way, won a battle many others had lost.

Before that, Denton residents had read and heard grim statistics about the rising AIDS death toll. But statistics are hard to visualize and so was AIDS, Acquired Immune Deficiency Syndrome. Hard to visualize, that is, until a little boy with the virus that can lead to AIDS arrived in Denton and wanted to go to the kindergarten.

A walk down Denton's brick sidewalks is a journey into small-town, front-doors-unlocked America. The Caroline County seat, Denton lies nestled among the corn fields, and chicken houses of Maryland's Eastern Shore, the peninsula east of the Chesapeake Bay. People are friendly but things move slowly here. Change comes slowly.

Maybe in a small town more than anywhere else, rumors are dangerous. Here in Denton, everyone knows everyone and all about everyone. And, if a little bit of knowledge is a dangerous thin, a little knowledge about a person with AIDS is the stuff nightmares are made of.

During May, 1987, word spread that a boy with AIDS had moved to the Denton area. Some parents discovered that the woman who babysat for their children also watched that little boy. The boy's mother had to find a new babysitter.

Still, things remained relatively quiet. The Caroline County Health Department held a meeting with concerned parents to answer their questions. The meeting started out emotionally but ended well, according to Caroline County Health Officer Dr. John A. Grant. "People just wanted to know and talked and talked and talked," Grant said. "We felt [the meeting] was very positive. As soon as they got information, they felt better."

But the possibility of a crisis still loomed and Grant embraced himself for the worst. He contacted other health officers to learn more about human immunodeficiency virus (HIV) and AIDS. He and others talked with the child's mother about starting over in another community where no one knew of her son's health problems.

The mother held firm. She and her son were staying in Denton. Although not obligated by state law, she let the school system know her son was infected with HIV. The little boy, a hemophiliac, had contracted it through a blood transfusion, the woman told officials. And, she said, she wanted her son to start kindergarten at Denton Elementary School in the fall.

Testing children – and policy

The mother's admission forced officials to test a year-old and as-yet-unused ADIS policy. State simply, the policy ordered that such a case be reviewed by a panel of school and health representatives to determine whether the child posed a threat to others. The panel used three standards set forth by the Academy

of Pediatrics. According to the academy, children with HIV should be allowed to attend school if they did not bleed uncontrollably, if they could control their excretory functions, and if they were not aggressive, biting children. The little boy in Denton met all three criteria.

No reason existed to bar the child from kindergarten. Grant told Superintendent of School William R. Ecker. Ecker agreed. Hysteria over a similar situation in Arcadia, Fla. worried him but the law and the school system policy were clear. The child would start school in September.

Word continued to spread. So did the rumors. A child with AIDS, or perhaps several children, were enrolled in Caroline County schools. Was it this school? No, it was that school. Who was this child?

And then at the Caroline County Board of Education meeting in August, a young woman rose to speak. She had heard something about AIDS. "What was the school system's policy on the issue?"

Board members listened and responded in broad terms. I was there as a reporter and took notes.

As the ashes smoldered in Arcadia, Fla., where fear of AIDS caused people to burn a family's home, things came to a head in Denton. In only a few days, schools in Caroline County would open and an unnamed little boy who so many people thought they knew so much about would start his first day of kindergarten.

"It was an explosive situation." Denton Elementary School Principle Charles Carey said months after the child's peaceful enrollment in his school.

He recalled his early caution with the press. When I first called him, for example, he kept his answers vague. This was a confidential matter and the child's identity had to be protected. But, a few minutes after he hung up from this conversation, he picked up the telephone again and called me back.

"I just went with a gut-level reaction" to trust the press, Carey told me. I had said I wanted to learn more about AIDS and Carey knew where I could get at least partially educated. He invited me to a faculty meeting called for later in the day to discuss AIDS.

After briefing teachers, Health Officer Grant told me that a child with HIV would be attending kindergarten in Caroline County in the fall. The story went to press and everyone held their breath.

The next day, the headline in the *Easton Star-Democrat* read: AIDS CHILD ENROLLED IN CAROLINE SCHOOL SAID NOT TO BE THREAT.

No one marched on Denton Elementary, Carey called a parents' meeting to be held shortly after school started where a panel, including Grant and a local doctor whose daughter was enrolled in the Denton elementary kindergarten, responded to parents' AIDS questions. Carol B. Seward, president of the elementary school P.T.A., was one of several parents who spoke in favor of the board's decision to admit the child.

The meeting was a typical Denton response to the AIDS experience. People were very concerned and very curious but remained cooperative and open-minded. Grant and other panel members fielded any and all questions and struck down rumors.

From then on, things have been calm in Denton save for visiting reporters wanting to get the story. The little boy at the center of the controversy finished his first year of school. Teachers and school nurses have grown accustomed to putting on rubber gloves when dealing with any child's wounds and other aseptic procedures are in place.

Just why Denton handled the AIDS crisis so well is still difficult to say. Certainly, Denton is filled with kind, loving people. But doesn't Arcadia, Fla. Have its share of good people, too?

Denton rose to the occasion, I believe, because of carefully considered openness. Rather than bow to emotion and panic, officials followed preset policies. Health and school leaders teamed up to answer each other's questions and those of the community.

Officials did not seek publicity but did not shun it either. They answered the questions of local residents and reporters. By getting out the facts, administrators were able to avert a crisis.

The local health department's involvement in all this was crucial. Doctors and nurses were consulted about every decision concerning the Denton kindergarten student. Their knowledge proved invaluable during meetings of the community. Educating the public about AIDS before and after summer, 1987 was also vital. No question was too trivial, although question-and-answer ground rules were made plain. The child's identity would not be given out. Instead, the facts about AIDS were the news for Caroline County.

Officials who turn down public requests for information are immediately viewed as having something to hide. In Caroline County, Carey, Ecker, and Grant thought before they spoke but they did, indeed, speak. By talking to the public through the press and meetings, they were able to get out their side and child's side of the story.

The courage and honesty of the child's mother must also be recognized in the Denton success story. She told school officials what they were facing rather than letting them find out for themselves. Forever anonymous, she spoke to the newspapers and let others know of her family's private sorrows. Readers sympathized and, perhaps, even learned.

Local newspapers also deserve credit in the Denton story. The *Caroline Times-Record and Easton Star-Democrat* ran stories and editorials full of AIDS background information.

It has been almost a year since AIDS grabbed Denton by the shoulders and shook it.

Pamela H. Payne-Foster

Awakened by a child to the reality of the disease, Denton learned what some still find difficult to grasp-that people are going to have to learn to live with AIDS even as it kills their neighbors.

GUIDELINES FOR AGE-APPROPRIATE AIDS PREVENTION EDUCATION

BASIC CONCEPTS K-12

1. AIDS is an infectious disease
2. AIDS is a preventable disease
3. There are things one can do to reduce the risk of infection
4. There are resources available for information and/or help regarding AIDS
5. Students are part of the network of resources to help address the problems surrounding AIDS
6. There are personal, societal, political, legal, and economic implications of AIDS
7. We owe it to ourselves and the larger society to actively work to prevent AIDS

SUGGESTED CONTENT BY GRADE LEVELS (samples)

K-3 AIDS is a serious disease affecting some adults
Very few children get AIDS
AIDS is hard to get; you don't catch it like you catch a cold or chicken pox
Be in charge of your own body; it's okay to say no to unwanted touch
If you have any questions, talk to your parents or an adult you trust

4-6 AIDS is a serious disease affecting some people
Nature of infectious diseases, viruses, immune system
Importance of good hygiene
How AIDS is not transmitted
Basic growth and development, including sexual development
Personal choices affect our health
Say no to drugs
If you have any questions, we can find the answers

7-8 AIDS is a serious disease, affecting young adults
How AIDS is transmitted (Anal, vaginal, oral intercourse, IV drug use with

shared needles, infected mother-child, etc.)
How AIDS is not transmitted
Delay of sexual activity
Avoidance of substance use/abuse
Avoidance of scapegoating, name-calling, rejection of persons with AIDS
Development of sense of social responsibility about AIDS prevention

9-12	AIDS is a serious disease that can be prevented How AIDS is not transmitted Abstinence from sexual intercourse and IV drug use is the best prevention If one is sexually active, use of condom reduces risk Encouragement of a waiting-is-worth it attitude toward sexual intercourse Skills in managing communication, dating, peer pressure, discussion of sexual choices beforehand, refusal skills, life planning, etc. Emphasis on both male and female responsibility for sexual behavior Consideration of moral/ethical aspects of relationships and sexual choices Awareness of political, legal, social, economic, health consequences

Skills of reaching out to others with compassion, caring
Emphasis on active involvement in HIV/AIDS prevention behaviors and advocacy

APPENDIX C

Additional Resources

Suggested Reading List
Books

1. AIDS Interfaith Network of Greater Harrisburg, PA. Developing AIDS Policies: A Manual for Congregations – Revs. Paul & Barbara Derrickson, September 1989, (717) 531-8177, (717) 782-5339.
2. The Black Church Speaks! A Collection of Historical Sermons on HIV/AIDS – Book and DVD. Found on www.balmingilead.org
3. My Rose- An African American Mother's Story of AIDS by Geneva E. Bell.
4. *Days of Grace* by Arthur Ashe.
5. *Daddy and Me: A Photo Story of Arthur Ashe and his daughter,*

Camera by Jeanne Moutoussamy-Ashe,

6. Oh God! A Black Woman's Guide to Sex and Spirituality by Reverend Dr. Susan Newman
7. The Secret Epidemic: The Story of AIDS and Black America by Jacob Levenson.
8. Living with HIV/AIDS: The Black Person's Guide to Prevention, Diagnosis and Treatment by Dr. Eric Goosby and Adrianne Appel
9. Beyond the Down Low: Sex, Lies, and Denial in Black America by Keith Boykin
10. The Boundaries of Blackness: AIDS and the Breakdown of Politics by Cathy J. Cohen
11. Out of Bounds by Roy Simmons and Damon DiMarco
12. Coming Up From the Down Low: The Journey to Acceptance and Honest Love by JL King and Courtney Carreras
13. On the Up and Up: A Survival Guide for Women Living with

Men on the Down Low by Brenda Stone Browder

14. African American Women and HIV/AIDS: Critical Responses, Edited by Dorie J. Gilbert and Ednita M. Wright
15. Social Workers Speak Out on the HIV/AIDS Crisis: Voices from African American Communities by Larry M. Grant, Patricia Stewart and Vincent J. Lynch with Willis Green, Jr., Darrell P. Wheeler and Ednita M. Wright
16. The Second Chapter: ACCEPTANCE by Shelton Jackson
17. A Pastor's Guidebook for HIV/AIDS – Ministry through the Church. Found on the www.arkofrefuge.org.

HIV/AIDS Links for African Americans – found on www.hivinsite.ucsf.edu/InSite

- African American AIDS Policy and Training Institute
  A program of the University of Southern California (USC) and its AIDS Social Policy Archives.

  - Kujisource
    Is African American AIDS Policy and Training Institute's (AAAPTI) monthly newsletter. Sign-up to receive your own free copy in the mail or check them out online.

- African-American HIV/AIDS Program of the American Red Cross
  Trains instructors to share community-based information about HIV, particularly with teens. Also develops posters and other prevention materials targeting African Americans.

- The Balm in Gilead

Mobilizes African American churches to respond to AIDS through conferences, radio programs, videos and community forums.

- The Black Church and HIV/AIDS
Web page with links to news, features, church statements, and worship resources related to African Americans, the church, and HIV/AIDS.

- Kaiser Family Foundation - HIV/AIDS
Includes Key Facts: African Americans and HIV/AIDS [PDF], and African Americans and HIV/AIDS Fact Sheet [PDF].

- National Black HIV/AIDS Awareness Day
A nationwide community mobilization effort to emphasize the HIV/AIDS

- National Black Nurses Association: HIV/AIDS Program
(NBNA) U.S.-based professional educational and advocacy organization of African American nurses.

- The 1999 National Conference on

African Americans and AIDS
A national forum for health professionals who provide care for African-Americans. Presentations in RealAudio format.

- National Medical Association: HIV/AIDS Program
(NMA) The largest and oldest national organization representing African American physicians and their patients in the United States.

- National Minority AIDS Council
An association of more than 3,000 AIDS service organizations that target people of color. NMAC can provide local referrals to member agencies upon request.

- Supporting Networks of HIV Care
A partnership between the CAEAR Coalition Foundation, the NMAC, and the HRSA HIV/AIDS Bureau that provides free assistance to nonprofit, community, and faith-based organizations with the development or improvement of their ability to provide primary health care and support

services to people of color living with or affected by HIV/AIDS.

Videos about African Americans and HIV/AIDS
From Video AIDS: A Catalog For Users of AIDS Educational Resources.

Other Resources:

- **Articles by Kai Wright including "Blacks' Plaque: A look at misconceptions about AIDS in the African American community" "When a Son is Gay" and "Secret Sex"**
- **Publications from the Black AIDS Institute including "The Time is Now! The State of AIDS in Black America" by Kai Wright and "Reclaiming Our Future: The State of AIDS Among Black Youth in America by Dr. Cathy J. Cohen, Alexandra Bell and Mosi Ifatunji**
- **African American HIV/AIDS Training materials by the American Red Cross including:**
- ***The Talking Drums (Dono Ntoaso)*. A leader's guide and workbook for building HIV prevention skills in African American communities**

- **_African American HIV Education and Prevention Instructor's Manual_, a tool for HIV/AIDS instructors to use in community-level HIV prevention education**
- **Don't Forget Sherrie Help young people learn and apply the facts about HIV and AIDS and develop effective decision-making skills (video, workbook, and leader's guide)**
- **Proverb Note cards –African Proverb note cards feature colorful, cultural images that interpret proverbs as messages about HIV and AIDS. Twelve cards, two of each poster design.**

About the Author

Dr. Pamela Payne Foster is a Preventive Medicine and Public Health physician who currently resides in Montgomery and Tuscaloosa, Alabama. She is Deputy Director of the Institute for Rural Health Research at The University of Alabama in Tuscaloosa,

Alabama. She has worked most of her career in the field of minority health through academic - community partnerships, serving both African-American and Latino communities within and near Long Island, New York, Atlanta, Georgia, Washington, D.C. and now Alabama. She is married to William Foster, Jr. a social worker. In 2006, they founded a nonprofit organization, AframSouth, Inc.; a culturally-based educational and health resource center focused on human and health development issues in African-American communities, particularly in the South.

*The National **AIDS Book Project** is designed to provide books free of charge to community members through the support of nonprofit organizations. For more information about the National **AIDS Book Project** please contact the Coordinator- Roland Barksdale-Hall at roland.barksdalehall@gmail.com*

Pamela H. Payne-Foster